matics and its applications

Editor: G. M. BELL, Professor of Mathematics, King' College (KQC), sity of London

atics and its applications are now awe-inspiring in their scope, variety and depth. ly is there rapid growth in pure mathematics and its applications to the traditional f the physical sciences, engineering and statistics, but new fields of application are g in biology, ecology and social organisation. The user of mathematics must assimi- otle new techniques and also learn to handle the great power of the computer efficiently onomically.

ed of clear, concise and authoritative texts is thus greater than ever and our series will our to supply this need. It aims to be comprehensive and yet flexible. Works survey- ent research will introduce new areas and up-to-date mathematical methods. Under- te texts on established topics will simulate student interest by including applications t at the present day. The series will also include selected volumes of lecture notes will enable certain important topics to be presented earlier than would otherwise be le.

these ways it is hoped to render a valuable service to those who learn, teach, develop e mathematics.

ck, G. R. & Bridgeman, T.	Mathematical Theory of Wave Motion
M. A.	Mathematics in the Social and Life Sciences: Theories, Models and Methods
rra, G.	Measure Theory and Integration
, J. S., Burghes, D. N., Huntley, I. D., James, D. J. G. & Moscardini, A. O.	Teaching and Applying Mathematical Modelling
hes, D. N. & Borrie, M.	Modelling with Differential Equations
hes, D. N. & Downs, A. M.	Modern Introduction to Classical Mechanics and Control
hes, D. N. & Graham, A.	Introduction to Control Theory, including Optimal Control
hes, D. N., Huntley, I. & McDonald, J.	Applying Mathematics
hes, D. N. & Wood, A. D.	Mathematical Models in the Social, Management and Life Sciences
ovskiy, A. G.	Green's Functions and Transfer Functions Handbook
ovskiy, A. G.	Structure of Distributed Systems
, Z-Q., Kim, K. H. & Roush, F. W.	Incline Algebra and Applications
rlton, F.	Textbook of Dynamics, 2nd Edition
rlton, F.	Vector and Tensor Methods
oper, G. D.	Introduction to Water Waves
ss, M. & Moscardini, A. O.	Learning the Art of Mathematical Modelling
len, M. R.	Linear Models in Biology
nning-Davies, J.	Mathematical Methods for Mathematicians, Physical Scientists and Engineers
on, G., Coles, C. W. & Gettinby, G.	Mathematics and Statistics for the Bio-sciences
on, H.	Handbook of Hypergeometric Integrals
on, H.	Multiple Hypergeometric Functions and Applications
on, H.	q-Hypergeometric Functions and Applications
ux, I. D. & Pratt, M. J.	Computational Geometry for Design and Manufacture
by, P. A. & Gardiner, C. F.	Surface Topology
rdiner, C. F.	Modern Algebra
sson, P. C.	Geometry of Spatial Forms
odbody, A. M.	Cartesian Tensors
ult, R. J.	Applied Linear Algebra
aham, A.	Kronecker Products and Matrix Calculus: with Applications
aham, A.	Matrix Theory and Applications for Engineers and Mathematicians
iffel, D. H.	Applied Functional Analysis

LIN

Professor of Ma

New

Mathe
Series
Univer

Mathen
Not o
fields
emergi
late su
and ec

The n
endea
ing re
gradu
releva
which
possi

In all
and u

Bald
Ball,
de B
Berr

Burg
Burg
Burg
Burg
Bur

But
But
Cao
Ch
Ch
Cra
Cr
Cu
Du

Ea
Ex
Ex
Ex
Fa
Fi
G
G
G
G
G
G
G

Harris, D. J.	Mathematics for Business, Management and Economics
Hanyga, A.	Mathematical Theory of Non-linear Elasticity
Hoskins, R. F.	Generalised Functions
Hunter, S. C.	Mechanics of Continuous Media, 2nd (Revised) Edition
Huntley, I. & Johnson, R. M.	Linear and Nonlinear Differential Equations
Jaswon, M. A. & Rose, M. A.	Crystal Symmetry: The Theory of Colour Crystallography
Johnson, R. M.	Linear Differential Equations and Difference Equations: A Systems Approach
Kim, K. H. & Roush, F. W.	Applied Abstract Algebra
Kosinski, W.	Field Singularities and Wave Analysis in Continuum Mechanics
Lord, E. A. & Wilson, C. B.	The Mathematical Description of Shape and Form
Marichev, O. I.	Integral Transforms of Higher Transcendental Functions
Meek, B. L. & Fairthorne, S.	Using Computers
Moore, R.	Computational Functional Analysis
Muller-Pfeiffer, E.	Spectral Theory of Ordinary Differential Operators
Nonweiller, T. R. F.	Computational Mathematics: An Introduction to Numerical Analysis
Oldknow, A. & Smith, D.	Learning Mathematics with Micros
Ogden, R. W.	Non-linear Elastic Deformations
Rankin, R.	Modular Forms
Ratschek, H. & Rokne, Jon	Computer Methods for the Range of Functions
Scorer, R. S.	Environmental Aerodynamics
Smith, D. K.	Network Optimisation Practice: A Computational Guide
Srivastava, H. M. & Karlsson, P. W.	Multiple Gaussian Hypergeometric Series
Srivastava, H. M. & Manocha, H. L.	A Treatise on Generating Functions
Sweet, M. V.	Algebra, Geometry and Trigonometry in Science, Engineering and Mathematics
Temperley, H. N. V. & Trevena, D. H.	Liquids and Their Properties
Temperley, H. N. V.	Graph Theory and Applications
Thom, R.	Mathematical Models of Morphogenesis
Townend, M. Stewart	Mathematics in Sport
Toth, G.	Harmonic and Minimal Maps
Twizell, E. H.	Computational Methods for Partial Differential Equations
Wheeler, R. F.	Rethinking Mathematical Concepts
Willmore, T. J.	Total Curvature in Riemannian Geometry
Willmore, T. J. & Hitchin, N.	Global Riemannian Geometry

Statistics and Operational Research

Editor: B. W. CONOLLY, Professor of Operational Research, Queen Mary College, University of London

Beaumont, G. P.	Introductory Applied Probability
Beaumont, G. P.	Basic Probability and Random Variables*
Conolly, B. W.	Techniques in Operational Research: Vol. 1, Queueing Systems
Conolly, B. W.	Techniques in Operational Research: Vol. 2, Models, Search, Randomization
French, S.	Sequencing and Scheduling: Mathematics of the Job Shop
French, S.	Decision Theory
Griffiths, P. & Hill, I. D.	Applied Statistics Algorithms
Hartley R.	Linear Methods of Mathematical Programming
Jones, A. J.	Game Theory
Kemp, K. W.	Dice, Data and Decisions: Introductory Statistics
Oliveira-Pinto, F.	Simulation Concepts in Mathematical Modelling
Oliveira-Pinto, F. & Conolly, B. W.	Applicable Mathematics of Non-physical Phenomena
Schendel, U.	Introduction to Numerical Methods for Parallel Computers
Stoodley, K. D. C.	Applied and Computational Statistics A First Course
Stoodley, K. D. C., Lewis, T. & Stainton, C. L. S.	Applied Statistical Techniques
Thomas, L. C.	Games, Theory and Applications
Whitehead, J. R.	The Design and Analysis of Sequential Clinical Trials

**In preparation*

First published in 1985 by
ELLIS HORWOOD LIMITED
Market Cross House, Cooper Street, Chichester, West Sussex, PO19 1EB, England

The publisher's colophon is reproduced from James Gillison's drawing of the ancient Market Cross, Chichester.

Distributors:

Australia, New Zealand, South-east Asia:
Jacaranda-Wiley Ltd., Jacaranda Press,
JOHN WILEY & SONS INC.,
G.P.O. Box 859, Brisbane, Queensland 4001, Australia

Canada:
JOHN WILEY & SONS CANADA LIMITED
22 Worcester Road, Rexdale, Ontario, Canada.

Europe, Africa:
JOHN WILEY & SONS LIMITED
Baffins Lane, Chichester, West Sussex, England.

North and South America and the rest of the world:
Halsted Press: a division of
JOHN WILEY & SONS
605 Third Avenue, New York, N.Y. 10158 U.S.A.

British Library Cataloguing in Publication Data
Cullen, M. R.
Linear models in biology. –
(Ellis Horwood series in mathematics and its applications)
1. Biology – Mathematical models
I. Title
574'.0724 QH323.5

Library of Congress Card No. 85–14045

ISBN 0–85312–835–9 (Ellis Horwood Limited – Library Edn.)
ISBN 0–85312–905–3 (Ellis Horwood Limited – Student Edn.)
ISBN 0–470–20205–X (Halsted Press – Library Edn.)
ISBN 0–470–20206–8 (Halsted Press – Student Edn.)

Printed in Great Britain by R.J. Acford, Chichester

Contents

To my teacher and colleague
Father Clarence J. Wallen, SJ

Introduction

Linear Models in Biology introduces the reader to a class of mathematical models that have proven to be useful in many areas of biology and medicine. The prerequisites are linear algebra, differential equations, and knowledge of a programming language. In addition, since all key theorems will be proved, the reader should have some exposure to mathematical proof. A good general background in biology is certainly helpful but not really necessary.

For the most part, the models that we will study are 'linear', that is, the mathematics needed to solve the equations of the model consists of what is ordinarily referred to as *linear systems analysis*. This includes linear algebra (and in particular eigenvalues and eigenvectors), systems of linear differential and difference equations, and linear optimization. This rather straightforward and well-developed body of mathematics should be familiar to most undergraduates by the end of their junior year. This allows such a student to concentrate on the models themselves. A disadvantage is that the models tend to be simplistic, but there are enough examples of linear models that have proven to be highly useful to offset this problem. Non-linear models in general require more sophisticated mathematical tools.

There are real advantages in presenting biological models versus models from the physical sciences. In the case of physics or engineering, models are constructed from the 'bottom-up', that is, model assumptions are based on rather well defined principles or laws, and it may not always be apparent what simplifications and compromises are being made. In biological modeling, there are few 'laws' to guide us and we may possess very limited knowledge about the system. In fact, the whole point of the model might be to discover hypotheses about the mechanisms for change that lead to predictions consistent with the known behavior. Such models are constructed from the 'top-down' and there is a great deal of freedom in selecting the model assumptions. It is precisely this freedom that makes biological models so ideal in a first course in modeling. The student can easily make alternate assumptions and explore their consequences. Most of the models do not require a deep knowledge of biology, and as a result it is easier to recognize the simplifications that have been made.

The modeling process is very much goal oriented. We may wish to develop a model that predicts drug concentrations in the bloodstream in terms of the initial drug dose. A model would consist of a set of reasonable assumptions about the pathways of the drug in the body and the rates at which the drug moves from place to place. In making these assumptions, we combine our knowledge of the chemical properties of the drug with guesswork.

$$\text{dose } x_0 \to \ldots ? \to \ldots ? \to \ldots \to \text{blood concentration } c(t)$$

In working through the models of the text, I could not help but be reminded of the wonderful machines invented by the American cartoonist and humorist Rube Goldberg. In the illustration below, the goal is to uncork the rather large bottle by a simple pull of the string. Rube has shown us an extremely imaginative and ingenious set of steps that take us to our final goal. Although we hardly have the wild and uninhibited freedom of a Rube Goldberg, we too must try to fill in the steps, guided by our present knowledge of the hidden system and good common sense. If our assumptions lead to conclusions that do not match the data, we must return to our model hypotheses and make alternate assumptions. The biological system is immensely more complicated than our model and so, in some sense, we will always be 'Rube Goldberging' it.

Our Own Self-Working Corkscrew, Rube Goldberg
(Reprinted with special permission of King Features Syndicate, Inc.)

Chapter 1 introduces the reader to discrete and continuous compartmental models and develops the matrix equations corresponding to these two cases. The eigenvalue–eigenvector solutions to these systems of linear difference or differential equations are developed in Chapter 2, while Chapter 3 treats the algorithms needed to implement these solution methods on the computer. In particular, the spectral decomposition theorem for matrices is proven and the associated power method for finding eigenvalues is presented.

The next three chapters treat discrete matrix models in population biology. The Leslie matrix model and the corresponding theorems on dominant eigenvalues are discussed in Chapter 4. The basic model is illustrated using several animal population studies from the literature. Chapter 5 presents several recent generalizations of the Leslie model including the two-sex model and Usher's forest harvesting model. Again examples and exercises are based on recent journal articles (e.g. the *Journal of Wildlife Management*). Linear programming methods are introduced in Chapter 6 to construct optimal harvesting models. Applications to forestry and wildlife management are presented.

The next five chapters concentrate on continuous linear models. Chapter 7 discusses methods for constructing solutions to non-homogeneous linear systems of the form $\dot{\mathbf{X}} = \mathbf{A}\,\mathbf{X} + \mathbf{F}(t)$ and emphasizes the matrix generalization of the method of undetermined coefficients.

Chapters 8 and 9 introduce the reader to tracer methods in physiology. The more elementary one- and two-compartment models are covered in Chapter 8. Applications include the dye-dilution method for measuring cardiac output, various glucose tolerance tests, Fick's principle, and pharmacokinetics. Chapter 9 examines the estimation of parameters problem in multi-compartmental models and solves the two-compartment case in detail. Applications include drug therapy models.

Chapter 10 presents general systems of differential equations, critical point analysis, and various methods (such as the Runge–Kutta techniques) for approximating solutions. The concepts are illustrated using classical predator–prey and competition models. Chapter 11 introduces the reader to systems ecology. After presenting the formulas needed for estimating transfer coefficients from field data, the chapter concludes with two case studies. The first study involves radioactive fall-out in a Puerto Rican rain forest, while the second study involves carbon cycling in the Aleut ecosystem.

Finally, Chapter 12 treats the important topic of sensitivity analysis. Formulas for state sensitivities to model parameters and eigenvalue sensitivity formulae are developed in detail and illustrated using some of the models that were presented earlier in the text.

A key part of this short text is the exercises. You will find about 200 exercises and many will require the computer for their solution. Part B exercises are more challenging and often extend the results in the text proper. You are encouraged to work as many of the exercises as possible.

I have purposely avoided any philosophical discussion of the modeling process itself. I suspect that such discourses are poorly understood by students until after they have labored through a large number of models themselves. Historically, many of the first attempts to model a given biological process have resulted in linear models. It therefore seems appropriate that an undergraduate student, equipped with the standard lower division courses in linear algebra and differential equations, begin with such models.

CHAPTER 1

Matrix Representation of Compartment Models

1.1 INTRODUCTION TO COMPARTMENT MODELS

In constructing a *compartment model* of a physical system, we first conceptually separate the system into a number of distinct *compartments* or *states* between which 'material' is transported. It is not necessary that the compartments be spatially distinct, but a clear set of rules must be given from which we can determine whether a given part of the system is contained in a compartment. This is illustrated in the following examples.

Example 1.1 The members of a population are divided into age classes and, as time goes on, they either move from age class to age class or pass out of the system entirely. A model that might be appropriate for a fish population is illustrated in Fig. 1.1.

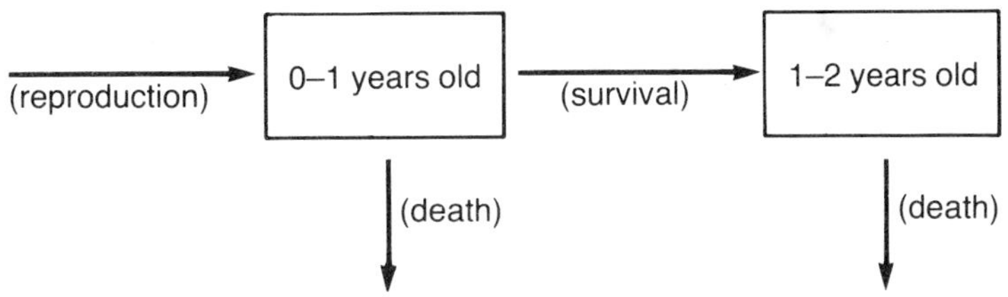

Fig. 1.1 Compartment diagram for a fish population (continued overleaf)

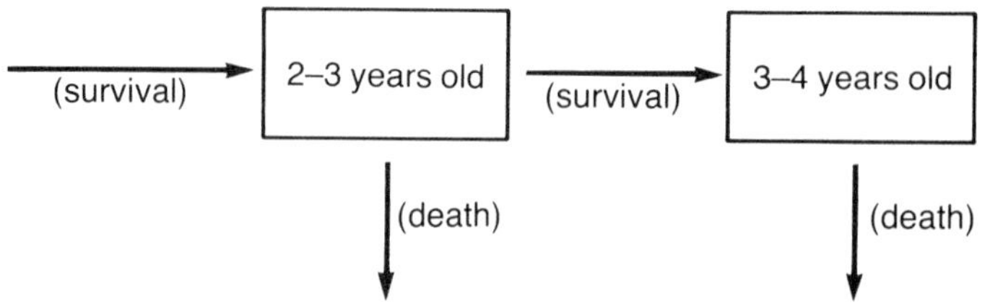

Fig. 1.1 (continued)

Example 1.2 A radioactive tracer (such as iodine-131) is injected into the bloodstream. Some of the tracer is metabolized by a particular organ (e.g. the thyroid gland) while the rest eventually makes its way to the kidneys where it is excreted. The model shown in Fig. 1.2 might be suitable. We could artificially create a fourth compartment labeled 'losses' to keep track of the amount of tracer lost to the system.

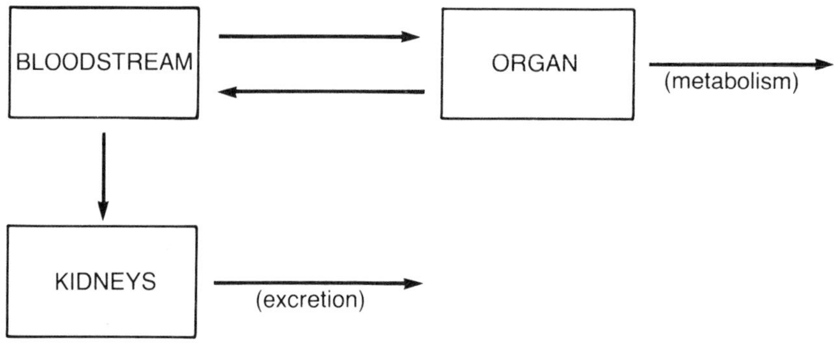

Fig. 1.2 Compartment diagram for a radioactive tracer

Example 1.3 Strontium-90 is deposited into pastureland by rainfall. To study how this material is cycled through the ecosystem, we might divide the system into the five compartments 'atmosphere', 'grasses', 'soil', 'dead organic matter', and 'streams' (Fig. 1.3). The modes of passing from one compartment to another (i.e. the *transport mechanisms*) are also shown.

Example 1.4 It is not necessary that the 'material' being transported be a physical substance. As an example, suppose that we classify the weather in a certain locale as 'rainy', 'overcast', or 'clear'. These are the three states of the system. We next keep track of the *probability* that the system is in a particular state at time t. Thus at a given moment the total probability of 1 will be distributed among the three comparments (Fig. 1.4).

On the arrow from compartment 1 to 2, for example, we place the *conditional probability* that it will rain tomorrow given that today is clear. Such probability models are known as *Markov chains*.

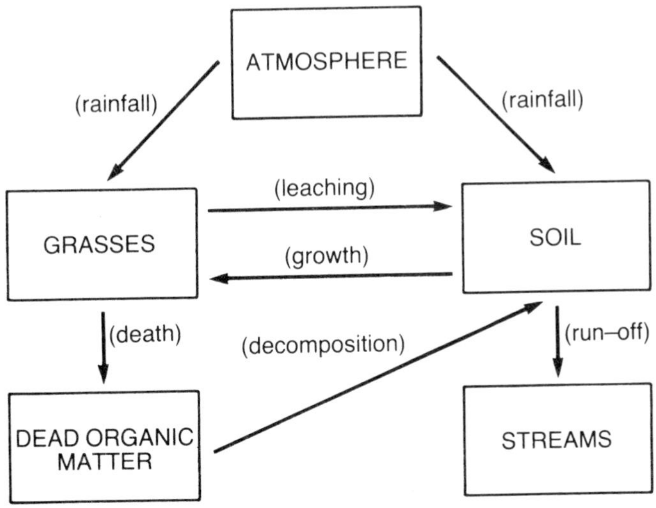

Fig. 1.3: Strontium 90 cycling in pastureland

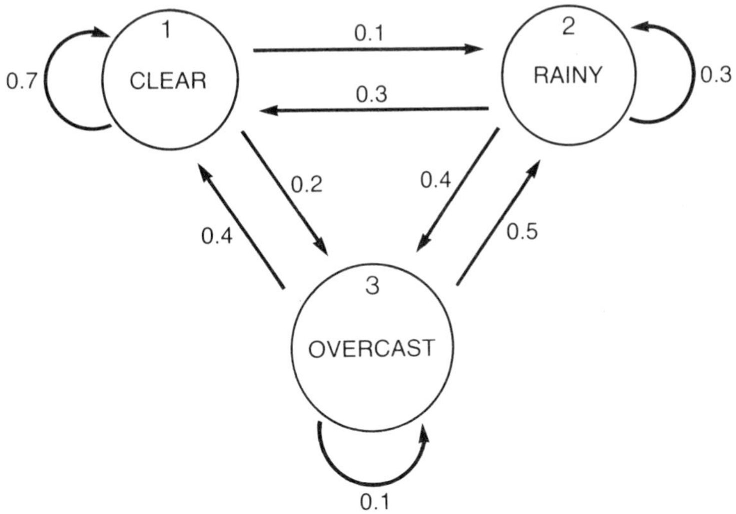

Fig. 1.4: Markov process for weather forecasting

1.2 TRANSFER COEFFICIENTS

Although the above examples show how varied the interpretations of 'compartment' and 'material' can be, as an aid in deriving the equations, let us imagine that a physical substance is being moved from box to box as shown in Fig. 1.5.

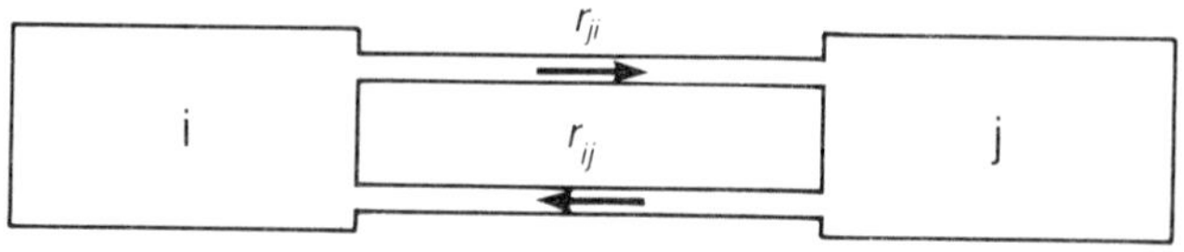

Fig. 1.5: Fluxes between two compartments

Let $x_i(t)$ = amount of material in compartment i at time t and form the vector $\mathbf{X}(t) = [x_1(t), x_2(t), \ldots, x_n(t)]$. *We will regard* $\mathbf{X}(t)$ *as either a row vector or a column vector*

$$\begin{bmatrix} x_1(t) \\ x_2(t) \\ \vdots \\ x_n(t) \end{bmatrix}$$

whichever is more convenient. $\mathbf{X}(t)$ then specifies the *state of the system* at time t.

Let $r_{ij}(t)$ be the rate at which material passes *to i from j* at a particular time t. Typical units for r_{ij} might be grams/minute. The rate function r_{ij} is also called the *flux* to compartment i from j. $r_{ij}(t)$ may be a function not only of t but also of the states $x_1, \ldots, x_n$. We will write, however, $r_{ij}(t)$ instead of $r_{ij}(x_1, x_2, \ldots, x_n, t)$ to keep the notation simple.

Finally, let $a_{ij}(t) = r_{ij}(t)/x_j(t)$. a_{ij} is called the *transfer coefficient* to i from j and is a percentage rate of change. Typical units for a_{ij} might be $(\text{minutes})^{-1}$. Thus if $a_{12} = 0.4$ per minute, then, at a given moment, material passes from 2 to 1 at the rate $0.4\,x_2(t)$ gm per minute.

The assumption that $a_{ij}(t) \equiv a_{ij}$ is constant for all i and j is known as the *linear donor-controlled hypothesis*. Thus $r_{ij}(t) = a_{ij}\,x_j(t)$ and the rate at which material passes to i from j is determined only by the donor $x_j(t)$. Although rather simplistic in nature, this assumption is quite common. In many cases there is simply not enough known data about the system to make alternate and more realistic assumptions. In any case, it is sound from a pedagogical point of view to start with these basic models and gradually increase the complexity.

Definition 1.1 If $a_{ik} = 0$ for all $i \neq k$, then compartment k is called a *sink* for the system. If $a_{kj} = 0$ for all $j \neq k$ we call compartment k a *source* for the system.

Thus, in Example 1.3, the atmosphere is a source while the compartment 'streams' is a sink. One can always create a sink for a compartment model by adding an $(n + 1)$st compartment called 'losses' which collects all materials leaving the system.

1.3 DISCRETE TRANSFERS

How can $\mathbf{X}(t)$ be found given the transfer coefficients a_{ij} and the initial state $\mathbf{X}(0)$? There are two different answers which depend on whether the material is *continually transported* between compartments or is *interchanged only at a discrete set of times* $t_1 = \Delta t,\ t_2 = 2\Delta t, \ldots, t_n = n\Delta t, \ldots$

If a_{ij} is the transfer coefficient and interchanges are made every Δt units, then at time $t_n + \Delta t$, $f_{ij} = a_{ij}\,\Delta t$ is the fraction of compartment j that is passed to compartment i. This is illustrated in the following example.

Example 1.5 A trust fund contains \$50 000, and transfers are to be made to a second account at the rate of 12% per year (Fig. 1.6). Thus, $a_{21} = 0.12$.

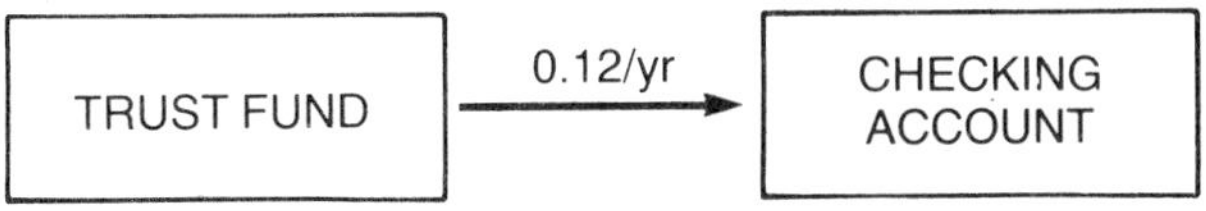

Fig. 1.6: Transfers into a checking account

If transfers are made monthly (i.e. $\Delta t = 1/12$), then $a_{21}\ \Delta t = 0.01$ or 1% is transferred each month. Thus $x_1(t + \Delta t) = x_1(t) - 0.01\,x_1(t) = 0.99\,x_1(t)$. If transfers are made daily (i.e. $\Delta t = 1/365$), then $a_{21}\ \Delta t = 0.12/365$ or 0.033% is transferred each day.

To develop the matrix equation relating $\mathbf{X}(t + \Delta t)$ to $\mathbf{X}(t)$, note that

$$
\begin{aligned}
x_i(t+\Delta t) &= x_i(t) + (\text{amount entering } i) - (\text{amount leaving } i)\\
&= x_i(t) + \sum_{j\neq i} f_{ij}\,x_j(t) - \sum_{j\neq i} f_{ji}\,x_i(t)\\
&= x_i(t) + \sum_{j\neq i} (a_{ij}\,\Delta t)\,x_j(t) - \left(\sum_{j\neq i} a_{ji}\right)\Delta t\,x_i(t).
\end{aligned}
$$

If we define $a_{ii} = -\sum_{j\neq i} a_{ji}$, we have

$$x_i(t+\Delta t) = x_i(t) + \Delta t \sum_{k=1}^{n} a_{ik}\,x_k(t).$$

Letting $\mathbf{A} = [a_{ij}]$, we have in matrix form $\mathbf{X}(t + \Delta t) = \mathbf{X}(t) + \Delta t\,\mathbf{A}\,\mathbf{X}(t)$ or

$$\boxed{\mathbf{X}(t+\Delta t) = (\mathbf{I} + \Delta t\,\mathbf{A})\mathbf{X}(t)}$$

where $\mathbf{I}$ is the $n \times n$ identity matrix. Note that the sum of the entries in any column of the matrix $\mathbf{A}$ is 0 since

$$a_{ii} + \sum_{j \neq i} a_{ji} = 0.$$

Such a matrix is called an *ecomatrix.*

Example 1.6 In the compartment diagram shown in Fig. 1.7, transfers are made each day.

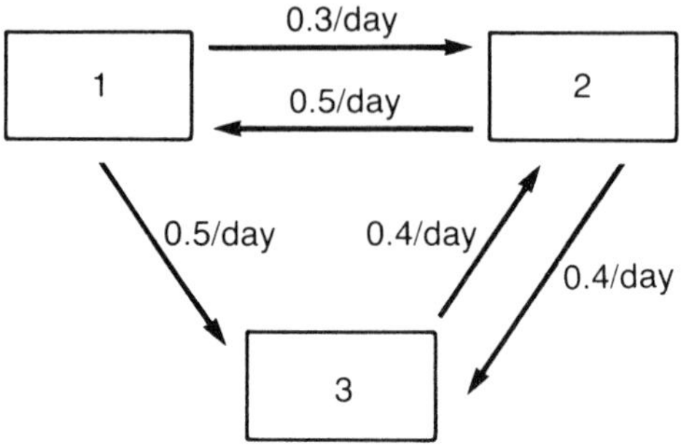

Fig. 1.7: Compartment diagram for example 1.6

(a) Find the transfer matrix **A**.
(b) If $\mathbf{X}(0) = [100, 0, 0]$, find the state of the system over the next five days.

Solution 1.6 We are given that $a_{21} = 0.3$, $a_{12} = 0.5$, $a_{32} = 0.4$, $a_{23} = 0.4$, $a_{13} = 0$, and $a_{31} = 0.5$. Since the sum of the entries in any column must be zero,

$$\mathbf{A} = \begin{bmatrix} -0.8 & 0.5 & 0 \\ 0.3 & -0.9 & 0.4 \\ 0.5 & 0.4 & -0.4 \end{bmatrix}$$

and for $\Delta t = 1$,

$$\mathbf{I} + \Delta t\ \mathbf{A} = \mathbf{I} + \mathbf{A} = \begin{bmatrix} 0.2 & 0.5 & 0 \\ 0.3 & 0.1 & 0.4 \\ 0.5 & 0.4 & 0.6 \end{bmatrix}$$

Hence $\mathbf{X}(1) = (\mathbf{I} + \mathbf{A})\ \mathbf{X}(0) = [20, 30, 50]$ and $\mathbf{X}(2) = (\mathbf{I} + \mathbf{A})\ \mathbf{X}(1) = [19, 29, 52]$. Continuing with the recursion $\mathbf{X}(t + 1) = (\mathbf{I} + \mathbf{A})\ \mathbf{X}(t)$, you should obtain $\mathbf{X}(3) = [18.3, 29.4, 52.3]$, $\mathbf{X}(4) = [18.36, 29.35, 52.29]$, and $\mathbf{X}(5) = [18.35, 29.36, 52.29]$.

If we let $\mathbf{B} = \mathbf{I} + \Delta t\ \mathbf{A}$, then $\mathbf{X}(t + \Delta t) = \mathbf{B}\ \mathbf{X}(t)$ and so $\mathbf{X}(\Delta t) = \mathbf{B}\ \mathbf{X}(0)$, $\mathbf{X}(2\ \Delta t) = \mathbf{B}\ \mathbf{X}(\Delta t) = \mathbf{B}(\mathbf{B}\ \mathbf{X}(0)) = \mathbf{B}^2\ \mathbf{X}(0)$, and in general $\mathbf{X}(n\ \Delta t) = \mathbf{B}^n\ \mathbf{X}(0)$. Hence:

$$\boxed{\mathbf{X}(n\,\Delta t) = (\mathbf{I} + \Delta t\ \mathbf{A})^n\ \mathbf{X}(0)}$$

This is the general solution for the discrete case.

1.4 CONTINUOUS TRANSFERS

By letting $\Delta t \to 0$ in the discrete case, we can derive the matrix differential equation for $\mathbf{X}(t)$ when the material flows continually between the compartments.

For Δt very small, $\mathbf{X}(t + \Delta t) \approx \mathbf{X}(t) + \Delta t\ \mathbf{A}\ \mathbf{X}(t)$ or $\mathbf{X}(t + \Delta t) - \mathbf{X}(t)/\Delta t \approx \mathbf{A}\ \mathbf{X}(t)$. Letting $\Delta t \to 0$, we obtain the system of differential equations

$$\boxed{\dot{\mathbf{X}}(t) = \mathbf{A}\ \mathbf{X}(t)}$$

where $\dot{\mathbf{X}}(t) = [\dot{x}_1(t), \dot{x}_2(t), \ldots, \dot{x}_n(t)]$.

Example 1.7 Water flows continuously between two 100 gallon tanks at the rate of 10 gallons/minute as shown in Fig. 1.8. Twenty pounds of salt are mixed

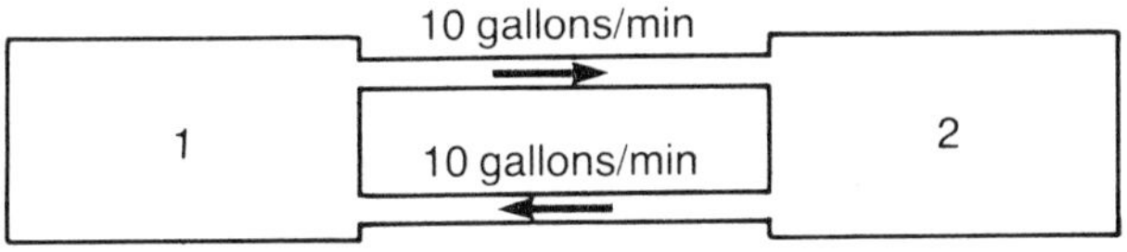

Fig. 1.8: Water flow between two tanks

in tank 1 at time $t = 0$. Since the concentration of salt in tank 1 at time t is $x_1(t)/100$ lb/gallon, salt moves to tank 2 at the rate

$$\begin{aligned} r_{21} &= (x_1(t)/100\ \text{lb/gallon})\ (10\ \text{gallons/min}) \\ &= 0.1\ x_1(t)\ \text{lb/min}. \end{aligned}$$

Thus $a_{21} = 0.1$. Likewise $a_{12} = 0.1$ and so

$$\mathbf{A} = \begin{bmatrix} -0.1 & 0.1 \\ 0.1 & -0.1 \end{bmatrix}.$$

Hence:

$$\begin{aligned} \dot{x}_1 &= -0.1\ x_1 + 0.1\ x_2 \\ \dot{x}_2 &= 0.1\ x_1 - 0.1\ x_2 \end{aligned}$$

with $x_1(0) = 20$ and $x_2(0) = 0$. This system may be solved by finding a second order differential equation for $x_1(t)$.

$$\begin{aligned}
\ddot{x}_1 &= -0.1\,\dot{x}_1 + 0.1\,\dot{x}_2 \\
&= -0.1\,\dot{x}_1 + 0.1\,(0.1\,x_1 - 0.1\,x_2), \text{ using the second equation} \\
&= -0.1\,\dot{x}_1 + 0.01\,x_1 - 0.1\,(0.1\,x_2) \\
&= -0.1\,\dot{x}_1 + 0.01\,x_1 - 0.1\,(\dot{x}_1 + 0.1\,x_1) \text{ using the first equation} \\
&= -0.2\,\dot{x}_1 .
\end{aligned}$$

Thus $x_1(t) = c_1 + c_2 e^{-0.2t}$ and $x_2(t) = 10[\dot{x}_1(t) + 0.1\,x_1(t)] = c_1 - c_2 e^{-0.2t}$. Since $x_1(0) = 20$ and $x_2(0) = 0$, it follows that $c_1 = 10$ and $c_2 = 10$. Thus $\lim_{t\to\infty} x_1(t) = \lim_{t\to\infty} x_2(t) = 10$, as expected.

To motivate the general form for the solution to $\dot{\mathbf{X}}(t) = \mathbf{A}\,\mathbf{X}(t)$, fix $t > 0$ and let $\Delta t = t/n$. Then, for n very large, $\mathbf{X}(t) = \mathbf{X}(n\,\Delta t)$ should be approximated by $(\mathbf{I} + \Delta t\;\mathbf{A})^n\;\mathbf{X}(0)$. Thus

$$\mathbf{X}(t) = (\mathbf{I} + \frac{1}{n}\;[t\mathbf{A}])^n\;\mathbf{X}(0) + \epsilon$$

where $\epsilon \to 0$ as $n \to +\infty$.

Recall that, for x a real number, $\lim_{n\to\infty} (1 + x/n)^n = e^x$. Our matrix expression resembles $(1 + ta/n)^n$ which approaches e^{ta}. If we define the matrix $e^{\mathbf{B}}$ to be $\lim_{n\to\infty} (\mathbf{I} + (1/n)\;\mathbf{B})^n$, then, assuming that the limit actually exists, we should have

$$\boxed{\mathbf{X}(t) = e^{t\mathbf{A}}\;\mathbf{X}(0)}$$

Actually, it is easier to define $e^{t\mathbf{A}}$ by the matrix series

$$e^{t\mathbf{A}} = \mathbf{I} + t\mathbf{A} + \frac{t^2}{2!}\;\mathbf{A}^2 + \frac{t^3}{3!}\;\mathbf{A}^3 + \ldots + \frac{t^n}{n!}\;\mathbf{A}^n + \ldots$$

and verify that $\mathbf{X}(t) = e^{t\mathbf{A}}\;\mathbf{X}(0)$ is a solution by direct differentiation. A rigorous development of matrix series and $e^{t\mathbf{A}}$ can be found in any analysis text that discusses normed linear spaces.

The explicit solutions $\mathbf{X}(n\;\Delta t) = (\mathbf{I} + \Delta t\;\mathbf{A})^n\;\mathbf{X}(0)$ and $\mathbf{X}(t) = e^{t\mathbf{A}}\;\mathbf{X}(0)$ arising from the discrete and continuous cases provide little insight into the long range behavior of the system. In the next chapter we will develop alternate forms for the solutions which are given in terms of the *eigenvalues* and *eigenvectors* of the matrix.

EXERCISES

Programming exercises

The programs requested in Exercises 1 and 2 should prove useful in doing many of the exercises that follow.

1 Use the recursion $\mathbf{X}(t + \Delta t) = \mathbf{B}\,\mathbf{X}(t)$ in writing a program that computes $\mathbf{X}(\Delta t), \mathbf{X}(2\,\Delta t), \ldots, \mathbf{X}(n\,\Delta t)$ given $\mathbf{X}(0)$.

2 (a) For a matrix A and real number h, write a program that finds

$$\mathbf{S}_n = \mathbf{I} + \mathbf{A}h + \frac{\mathbf{A}^2 h^2}{2} + \ldots + \frac{\mathbf{A}^n h^n}{n!}.$$

(b) For the differential equation $\dot{\mathbf{X}} = \mathbf{AX}$, show that $\mathbf{X}(t + h) = e^{\mathbf{A}h}\,\mathbf{X}(t)$. (*Hint*: Use the fact that $e^{\mathbf{A}(s+t)} = e^{\mathbf{A}s}\,e^{\mathbf{A}t}$.)

(c) Use (a) and the recursion in (b) to estimate $\mathbf{X}(h), \mathbf{X}(2h), \ldots, \mathbf{X}(nh)$ given $\mathbf{X}(0)$.

Part A

3 For the compartment diagram shown below,

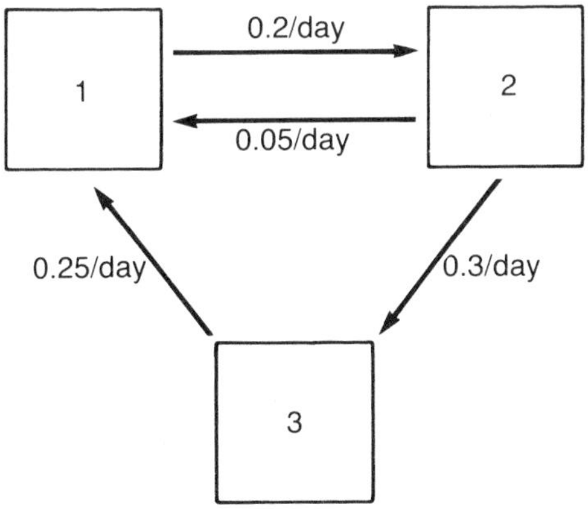

(a) Find the transfer matrix **A**.
(b) If $\mathbf{X}(0) = [100, 250, 80]$, find $\mathbf{X}(1)$ if material is transferred once a day.
(c) Find $\mathbf{X}(1)$ if material is transferred hourly.
(d) Estimate $\mathbf{X}(1)$ if material is transferred continuously.

4 For the compartment diagram below,

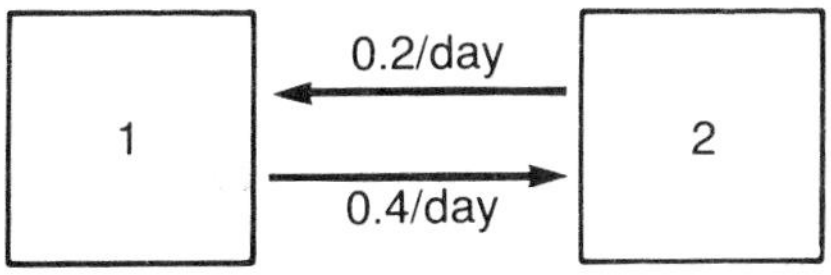

(a) Find the transfer matrix **A**.
(b) If $\mathbf{X}(0) = [90, 60]$, compute $\mathbf{X}(1), \ldots, \mathbf{X}(10)$ assuming that transfers are made once a day.
(c) Find the equilibrium state $\hat{\mathbf{X}}$ by solving $(\mathbf{I} + \mathbf{A})\mathbf{X} = \mathbf{X}$. (*Hint*: $x_1 + x_2 = 150$.)

5 Radioisotopes (such as phosphorus-32 and carbon-14) have been used to study the transfer of nutrients in food chains. Shown in the figure is a compartmental representation of a simple aquatic food chain. One hundred units (e.g., microcuries) of tracer are dissolved in the water of an aquarium containing a species of phytoplankton and a species of zooplankton.

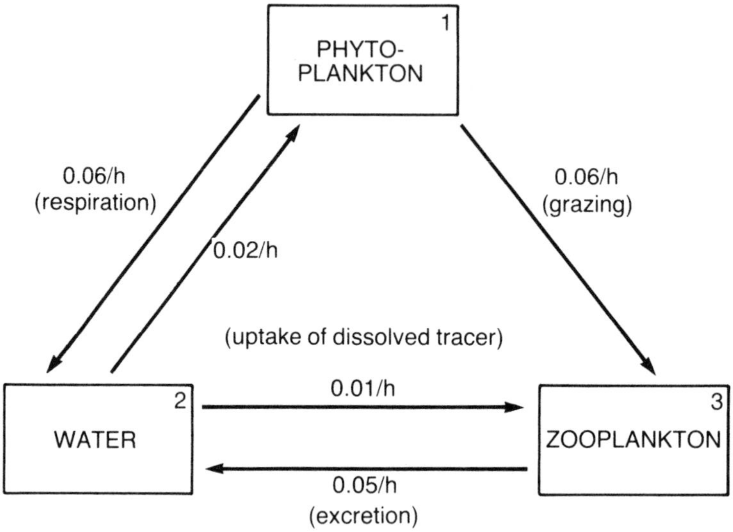

(a) Find the transfer matrix **A**.
(b) Assuming continuous transfers, estimate $\mathbf{X}(1), \ldots, \mathbf{X}(6)$.
(c) Find the equilibrium state $\hat{\mathbf{X}}$. (*Hint*: Solve $\mathbf{AX} = \mathbf{0}$ with $x_1 + x_2 + x_3 = 100$.)

(Adapted from M. R. Cullen, *Mathematics for the Biosciences*, PWS Publishers, 1983.)

6 A field has been completely devastated by fire. Two types of vegetation, grasses and small shrubs, will first begin to grow, but the small shrubs can take over an area only if preceded by the grasses. In the accompanying figure, the transfer coefficient of 0.3 indicates that, by the end of the summer, 30% of the prior bare space in the field becomes occupied by grasses.

(a) Find the transfer matrix **A**.
(b) If $\mathbf{X}(0) = [10, 0, 0]$ with area measured in acres, predict the ground cover over the next ten years.

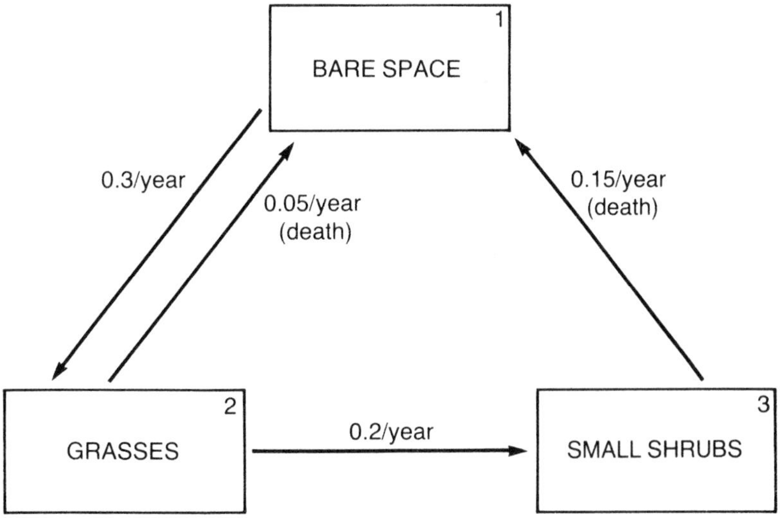

(Adapted from M. R. Cullen, *Mathematics for the Biosciences*, PWS Publishers, 1983).

7 The compartment model below illustrates the cycling of DDT when crops are sprayed.

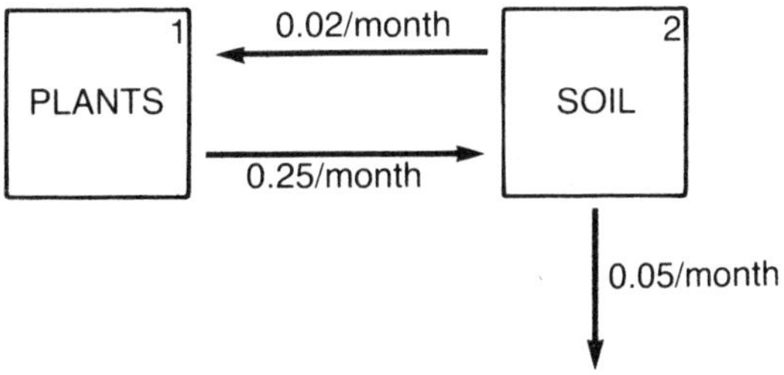

(a) Find the transfer matrix **A**.
(b) Solve $\dot{\mathbf{X}} = \mathbf{A}\,\mathbf{X}$ subject to $\mathbf{X}(0) = [0.6, 0.4]$ by first finding a second order differential equation satisfied by $x_1(t)$.

8 A two-tank system is shown in the figure on page 24. If 50 grams of dye are deposited into tank 1, find the expressions for $x_1(t)$ and $x_2(t)$.
(*Hint*: Find a second order differential equation satisfied by $x_1(t)$.)

(Adapted from M. R. Cullen, *Mathematics for the Biosciences*, PWS Publishers, 1983.)

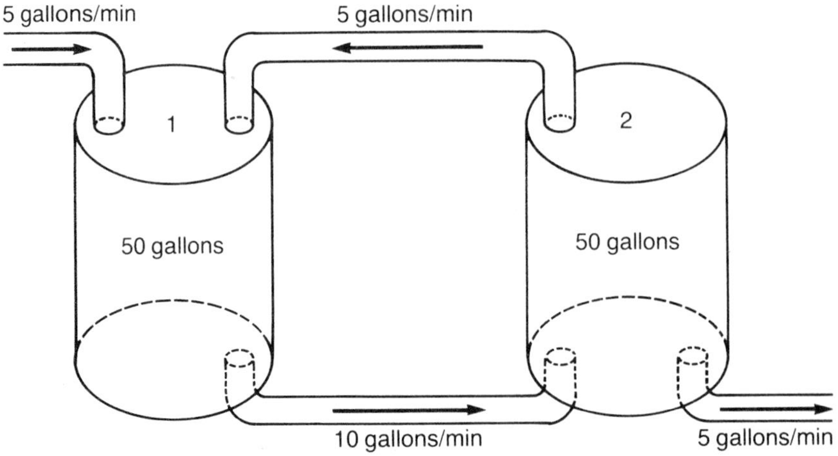

9 Shown below is the compartmental diagram for cesium-137 transfer in an Arctic food chain given by Eberhardt and Hanson [1]

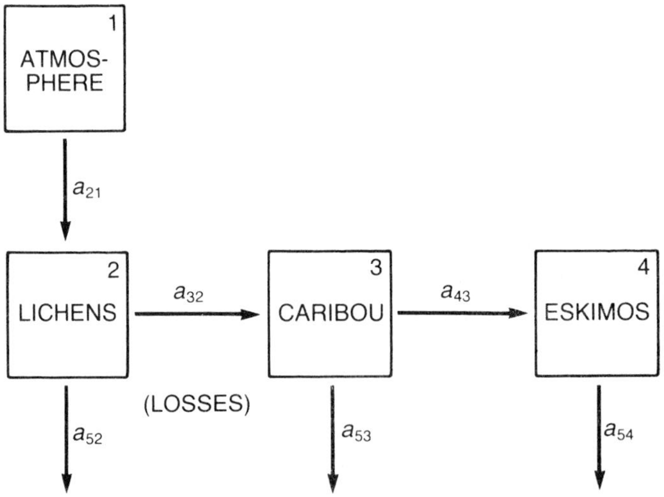

(a) Find $x_1(t)$ given that $x_1(0) = a$.
(b) Find $x_2(t)$ given that $x_2(0) = 0$.

10 The matrix **A** given below is the transfer matrix for biomass transfer in an oak-pine forest [2]:

$$\mathbf{A} = \begin{bmatrix} -0.215 & 0 & 0 & 0 \\ 0.07 & -0.437 & 0 & 0 \\ 0 & 0.262 & -0.0787 & 0 \\ 0.145 & 0.175 & 0.0787 & 0 \end{bmatrix}$$

The compartments are (1) vegetation, (2) litter, (3) humus, and (4) losses. Reconstruct the compartment diagram.

11 The matrix given below is the transfer matrix for energy dynamics in a tropical rain forest [3]:

$$\mathbf{A} = \begin{bmatrix} -0.0366 & 0 & 0 & 0 & 0 \\ 0.000146 & -0.00589 & 0 & 0 & 0 \\ 0.0000124 & 0.00112 & -0.0721 & 0 & 0 \\ 0 & 0.00262 & 0 & -0.000107 & 0 \\ 0.0364 & 0.00214 & 0.0721 & 0.000107 & 0 \end{bmatrix}$$

The compartments are (1) vegetation, (2) litter, (3) animals, (4) soil, and (5) respiration. Reconstruct the compartment diagram.

Part B

12 If $\mathbf{A}$ is an ecomatrix, show that $\det \mathbf{A} = 0$.

13 An ecomatrix is called *irreducible* in case it is possible for material to move from one compartment to any other compartment in the system. Thus the transfer matrix for the model below is irreducible even though there is no direct way for material to pass from (1) to (3).

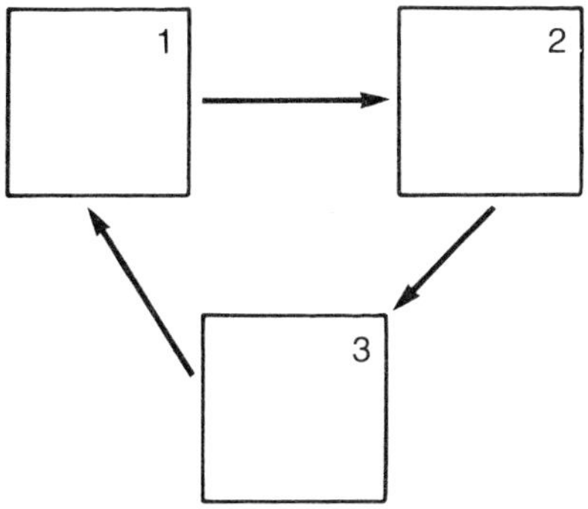

Which of the following transfer matrices are irreducible?

(a)

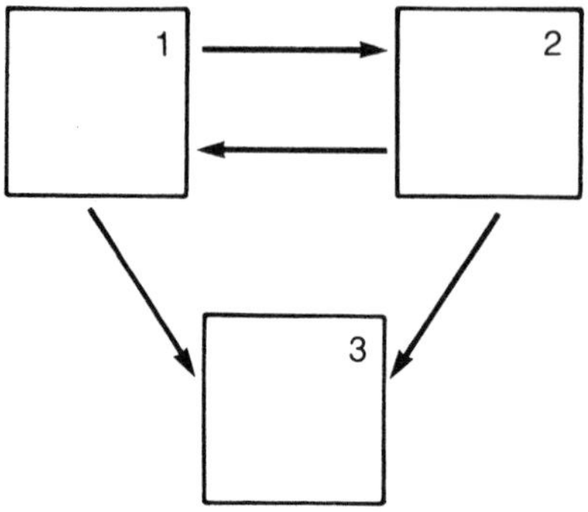

(b)

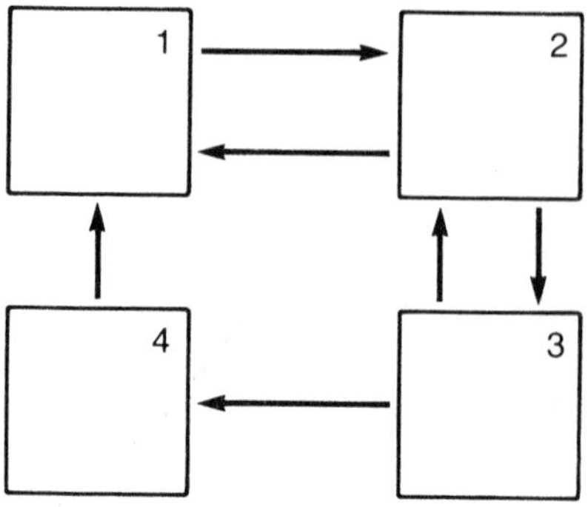

REFERENCES

[1] L. Eberhardt and W. Hanson, 'A Simulation Model for an Arctic Food Chain', *Health Physics* (1969), **17**, 793–806.

[2] H. Odum, *A Tropical Rain Forest*, U.S. Atomic Energy Commission, 1970.

[3] G. M. Woodwell, 'The Energy Cycle of the Biosphere', *Scientific American* (1970), **223**(3), 64–74.

CHAPTER 2

Solutions by Eigenvalue Analysis

2.1 EIGENVALUE–EIGENVECTOR SOLUTIONS

If $\mathbf{A}$ is an $n \times n$ matrix, the *eigenvalues* of $\mathbf{A}$ are the solutions to the polynomial equation

$$p(\lambda) = \det(\lambda \mathbf{I} - \mathbf{A}) = 0.$$

Recall that when $\det(\lambda \mathbf{I} - \mathbf{A}) = 0$, the matrix $\lambda \mathbf{I} - \mathbf{A}$ is singular and so the equation $(\lambda \mathbf{I} - \mathbf{A})\mathbf{X} = \mathbf{0}$ has at least one non-zero solution $\mathbf{E}_\lambda$. The vector $\mathbf{E}_\lambda$ is called an *eigenvector* corresponding to λ. Thus

$$\mathbf{A}\, \mathbf{E}_\lambda = \lambda\, \mathbf{E}_\lambda$$

Note that for any non-zero scalar α, $\alpha\, \mathbf{E}_\lambda$ is again an eigenvector since $\mathbf{A}(\alpha\, \mathbf{E}_\lambda) = \alpha\, \mathbf{A}\, \mathbf{E}_\lambda = \alpha\, \lambda\, \mathbf{E}_\lambda = \lambda(\alpha\, \mathbf{E}_\lambda)$.

Once an eigenvalue and corresponding eigenvector are found, we can construct solutions $\mathbf{X}(t)$ to either $\dot{\mathbf{X}} = \mathbf{A}\, \mathbf{X}$ or $\mathbf{X}(t + \Delta t) = \mathbf{A}\, \mathbf{X}(t)$. This is the content of our next two theorems.

Theorem 2.1 Suppose that λ is an eigenvalue of $\mathbf{A}$ and $\mathbf{E}_\lambda$ is a corresponding eigenvector. Then $\mathbf{X}(t) = e^{\lambda t}\, \mathbf{E}_\lambda$ is a solution to the differential equation $\dot{\mathbf{X}} = \mathbf{A}\, \mathbf{X}$.

Proof of Theorem 2.1 $\dot{\mathbf{X}}(t)$ is just $\lambda\, e^{\lambda t}\, \mathbf{E}_\lambda = e^{\lambda t}\, (\lambda\, \mathbf{E}_\lambda)$. Since $\lambda\, \mathbf{E}_\lambda = \mathbf{A}\, \mathbf{E}_\lambda$, we have

$$\dot{\mathbf{X}}(t) = e^{\lambda t}\, (\mathbf{A}\, \mathbf{E}_\lambda) = \mathbf{A}(e^{\lambda t}\, \mathbf{E}_\lambda) = \mathbf{A}\, \mathbf{X}(\mathrm{t}).$$

Theorem 2.2 Suppose that λ is an eigenvalue of $\mathbf{A}$ and $\mathbf{E}_\lambda$ is a corresponding eigenvector. Then $\mathbf{X}(t) = \lambda^{t/\Delta t}\ \mathbf{E}_\lambda$ is a solution to $\mathbf{X}(t + \Delta t) = \mathbf{A}\ \mathbf{X}(t)$ for $t = 0, \Delta t, 2\,\Delta t, \ldots, n\,\Delta t, \ldots$

Proof of Theorem 2.2 $\mathbf{X}(t + \Delta t) = \lambda^{(t+\Delta t)/\Delta t}\ \mathbf{E}_\lambda = \lambda^{t/\Delta t}(\lambda \mathbf{E}_\lambda)$. Since $\mathbf{A}\ \mathbf{E}_\lambda = \lambda\ \mathbf{E}_\lambda$, we have

$$\mathbf{X}(t + \Delta t) = \lambda^{t/\Delta t}\ \mathbf{A}\ \mathbf{E}_\lambda = \mathbf{A}\ (\lambda^{t/\Delta t}\ \mathbf{E}_\lambda) = \mathbf{A}\ \mathbf{X}(t).$$

Once we have found solutions $\mathbf{X}_1(t), \mathbf{X}_2(t), \ldots, \mathbf{X}_k(t)$, then the linear combination $\mathbf{X}(t) = c_1\ \mathbf{X}_1(t) + c_2\ \mathbf{X}_2(t) + \ldots + c_k\ \mathbf{X}_k(t)$ is again a solution. For the differential equation $\dot{\mathbf{X}} = \mathbf{A}\ \mathbf{X}$, this follows from the computation

$$\dot{\mathbf{X}}(t) = \sum_{i=1}^{k} c_i\ \dot{\mathbf{X}}_i(t) = \sum_{i=1}^{k} c_i\ \mathbf{A}\ \mathbf{X}_i(t)$$

$$= \mathbf{A}\left(\sum_{i=1}^{k} c_i\ \mathbf{X}_i(t)\right) = \mathbf{A}\ \mathbf{X}(t).$$

Example 2.1 Construct solutions to the differential equation $\dot{\mathbf{X}} = \mathbf{A}\ \mathbf{X}$ if $\mathbf{A}$ is the matrix

$$\mathbf{A} = \begin{bmatrix} 1 & 4 \\ 1 & 1 \end{bmatrix}$$

Solution 2.1 $p(\lambda) = \det(\lambda \mathbf{I} - \mathbf{A}) = \lambda^2 - 2\lambda - 3 = (\lambda - 3)(\lambda + 1)$. Therefore the eigenvalues are 3 and -1. To find an eigenvector corresponding to 3, we must solve $\mathbf{A}\ \mathbf{X} = 3\ \mathbf{X}$, that is,

$$x_1 + 4x_2 = 3x_1$$

$$x_1 + x_2 = 3x_2$$

Hence $x_1 = 2x_2$. Setting $x_2 = 1$, $[2, 1]$ is an eigenvector. Likewise $[-2, 1]$ is an eigenvector for $\lambda = -1$. By Theorem 2.1, both $\mathbf{X}_1(t) = e^{3t}\ [2, 1]$ and $\mathbf{X}_2(t) = e^{-t}\ [-2, 1]$ are solutions, and, for any choice of constants c_1 and c_2, $c_1\ \mathbf{X}_1(t) + c_2\ \mathbf{X}_2(t)$ is a solution.

2.2 SOLVING INITIAL VALUE PROBLEMS

How do we know whether we have found *all solutions*? We next show that once we have found *n linearly independent eigenvectors for* $\mathbf{A}$, then we can solve any initial value problem $\mathbf{X}(0) = \mathbf{X}_0$. Theorem 2.3 shows that distinct eigenvalues give rise to linearly independent eigenvectors.

Theorem 2.3 If $\mathbf{E}_1, \mathbf{E}_2, \ldots, \mathbf{E}_k$ are eigenvectors corresponding to distinct eigenvalues $\lambda_1, \lambda_2, \ldots, \lambda_k$, then vectors $\mathbf{E}_1, \mathbf{E}_2, \ldots, \mathbf{E}_k$ are linearly independent.

Proof of Theorem 2.3 The proof is by induction on the number of eigenvalues. Assume that the theorem is true for k, and let

$$\mathbf{E}_1, \mathbf{E}_2, \ldots, \mathbf{E}_k, \mathbf{E}_{k+1}$$

be the eigenvectors corresponding to distinct eigenvalues $\lambda_1, \ldots, \lambda_{k+1}$. Next assume that

$$\sum_{i=1}^{k+1} c_i \mathbf{E}_i = \mathbf{0}.$$

We must show that $c_i = 0$ for $i = 1, \ldots, k+1$. Multiplying by λ_{k+1} we have

$$\mathbf{0} = \sum_{i=1}^{k+1} c_i \lambda_{k+1} \mathbf{E}_i = \sum_{i=1}^{k} c_i \lambda_{k+1} \mathbf{E}_i + c_{k+1} \lambda_{k+1} \mathbf{E}_{k+1}$$

Multiplying by the matrix $\mathbf{A}$,

$$\mathbf{0} = \sum_{i=1}^{k+1} \mathbf{A}(c_i \mathbf{E}_i) = \sum_{i=1}^{k+1} c_i \lambda_i \mathbf{E}_i = \sum_{i=1}^{k} c_i \lambda_i \mathbf{E}_i + c_{k+1} \lambda_{k+1} \mathbf{E}_{k+1}$$

Subtracting the above two equations, we have

$$\sum_{i=1}^{k} c_i(\lambda_{k+1} - \lambda_i)\mathbf{E}_i = \mathbf{0}.$$

Since by induction, $\mathbf{E}_1, \ldots, \mathbf{E}_k$ are linearly independent, $c_i(\lambda_{k+1} - \lambda_i) = 0$ for $i = 1, \ldots, k$. Since the eigenvalues are distinct, $c_i = 0$ for $i = 1, \ldots, k$. Thus $\mathbf{0} = \sum_{i=1}^{k+1} c_i \mathbf{E}_i = c_{k+1} \mathbf{E}_{k+1}$. Since $\mathbf{E}_{k+1}$ is non-zero, $c_{k+1} = 0$ and the induction is complete.

Therefore, if matrix $\mathbf{A}$ has n distinct eigenvalues, we may find n linearly independent eigenvectors. $\mathbf{A}$ need not have n distinct eigenvalues. In fact, in many of our subsequent applications, you will see that $\lambda = 0$ is often a multiple root of $p(\lambda)$. For most matrices that arise from applied problems, however, n linearly independent eigenvectors can be found. We will first examine this important special case.

If $\mathbf{E}_1, \mathbf{E}_2, \ldots, \mathbf{E}_n$ are linearly independent eigenvectors, let $\mathbf{Z}$ be the matrix formed by using $\mathbf{E}_k$ as the kth column of $\mathbf{Z}$:

$$\mathbf{Z} = \left[\mathbf{E}_1 \,\vdots\, \mathbf{E}_2 \,\vdots\, \ldots \,\vdots\, \mathbf{E}_n\right].$$

Since the columns of $\mathbf{Z}$ are linearly independent, $\mathbf{Z}$ is *invertible.*

Let $\mathbf{X}(t)$ be either $\sum_{k=1}^{n} c_k\, e^{\lambda_k t}\, \mathbf{E}_k$ or $\sum_{k=1}^{n} c_k\, (\lambda_k)^{t/\Delta t}\, \mathbf{E}_k$. The constants constants $c_1, \ldots, c_n$ can be computed in terms of $\mathbf{X}(0)$ and $\mathbf{Z}$ as follows:

$$\begin{aligned}
\mathbf{X}(0) &= \sum_{k=1}^{n} c_k\, \mathbf{E}_k \\
&= \left[\mathbf{E}_1 \,\vdots\, \mathbf{E}_2 \,\vdots \ldots \vdots\, \mathbf{E}_n\right] \begin{bmatrix} c_1 \\ c_2 \\ \vdots \\ c_n \end{bmatrix} \\
&= \mathbf{Z}\,\mathbf{C}
\end{aligned}$$

Since $\mathbf{Z}$ is invertible, $\mathbf{C} = \mathbf{Z}^{-1}\, \mathbf{X}(0)$.

Example 2.2 Solve the difference equation $\mathbf{X}(t + 1) = \mathbf{A}\, \mathbf{X}(t)$ subject to $\mathbf{X}(0) = [2, 3]$, where $\mathbf{A}$ is the matrix in Example 2.1.

Solution 2.2 The matrix $\mathbf{Z} = \begin{bmatrix} 2 & -2 \\ 1 & 1 \end{bmatrix}$ of eigenvectors has inverse

$$\mathbf{Z}^{-1} = \begin{bmatrix} 1/4 & 1/2 \\ -1/4 & 1/2 \end{bmatrix}$$

and so $\mathbf{C} = \mathbf{Z}^{-1}\, \mathbf{X}(0) = [2, 1]$. Hence, $c_1 = 2$ and $c_2 = 1$, and so

$$\mathbf{X}(t) = 2\; 3^t\, [2, 1] + (-1)^t\, [-2, 1]$$

for $t = 0, 1, \ldots, n, \ldots$

Example 2.3 Find the solution to the differential equation $\dot{\mathbf{X}} = \mathbf{A}\,\mathbf{X}$ subject to $\mathbf{X}(0) = [0, 2, 1]$, where

$$\mathbf{A} = \begin{bmatrix} 4 & -2 & 1 \\ -2 & 1 & 2 \\ 1 & 2 & 4 \end{bmatrix}$$

Solution 2.3 The characteristic polynomial $p(\lambda) = \det(\lambda \mathbf{I} - \mathbf{A})$ is easily seen to be $(\lambda - 5)^2\, (\lambda + 1)$. Since $\lambda = 5$ is a multiple root, Theorem 2.3 does *not* guarantee that there will be three linearly independent eigenvectors. To find eigenvectors corresponding to 5, we must solve $\mathbf{A}\,\mathbf{X} = 5\,\mathbf{X}$ or $(\mathbf{A} - 5\,\mathbf{I})\mathbf{X} = \mathbf{0}$. This gives

$$-x_1 - 2x_2 + x_3 = 0$$
$$-2x_1 - 4x_2 + 2x_3 = 0$$
$$x_1 + 2x_2 - x_3 = 0$$

All three equations reduce to $x_3 = x_1 + 2x_2$, and x_1 and x_2 are arbitrary. If $x_1 = 0$ and $x_2 = 1$, then $x_3 = 2$. When $x_1 = 1$ and $x_2 = 0$, then $x_3 = 1$. Hence [0, 1, 2] and [1, 0, 1] are eigenvectors corresponding to $\lambda = 5$. An eigenvector corresponding to $\lambda = -1$ is $[-1, -2, 1]$. Thus

$$\mathbf{Z} = \begin{bmatrix} 0 & 1 & -1 \\ 1 & 0 & -2 \\ 2 & 1 & 1 \end{bmatrix} \quad \text{and} \quad \mathbf{Z}^{-1} = \frac{1}{6}\begin{bmatrix} -2 & 2 & 2 \\ 5 & -2 & 1 \\ -1 & -2 & 1 \end{bmatrix}.$$

It follows that $\mathbf{C} = \mathbf{Z}^{-1}\,\mathbf{X}(0) = [1, -1/2, -1/2]$ and so

$$\mathbf{X}(t) = e^{5t}[0, 1, 2] - \frac{1}{2}e^{5t}[1, 0, 1] - \frac{1}{2}e^{-t}[-1, -2, 1]$$
$$= e^{5t}[-1/2, 1, 3/2] + e^{-t}[1/2, 1, -1/2]$$

Note that for large t, $\mathbf{X}(t) \approx e^{5t}[-1/2, 1, 3/2]$.

The steps for solving $\dot{\mathbf{X}} = \mathbf{A}\,\mathbf{X}$ or $\mathbf{X}(t + \Delta t) = \mathbf{A}\,\mathbf{X}(t)$ by the eigenvalue method can be summarized as follows:

Step 1: Given $\mathbf{A}$, compute $p(\lambda) = \det(\lambda\mathbf{I} - \mathbf{A})$
Step 2: Find all roots (real and complex) to $p(\lambda) = 0$.
Step 3: Given eigenvalues $\lambda_1, \ldots, \lambda_n$, try to find n linearly eigenvectors $\mathbf{E}_1, \ldots, \mathbf{E}_n$.
Step 4: Let $\mathbf{Z} = \left[\mathbf{E}_1 \vdots \mathbf{E}_2 \vdots \ldots \vdots \mathbf{E}_n\right]$, and compute $\mathbf{Z}^{-1}$.
Step 5: Given the initial condition $\mathbf{X}(0)$, let $\mathbf{C} = \mathbf{Z}^{-1}\,\mathbf{X}(0)$.
Step 6: (a) Form $\mathbf{X}(t) = \sum_{k=1}^{n} c_k\,\mathbf{E}_k\,e^{\lambda_k t}$ for the differential equation.
(b) Form $\mathbf{X}(t) = \sum_{k=1}^{n} c_k\,\mathbf{E}_k(\lambda_k)^{t/\Delta t}$ for the difference equation.

The eigenvalue solution is useful in determining the long range behavior of $\mathbf{X}(t)$. Suppose that all eigenvalues are real, and that

$$\lambda_1 > \lambda_2 \geqslant \lambda_3 \geqslant \ldots \geqslant \lambda_n$$

λ_1 is called the *crucial eigenvalue* of $\mathbf{A}$. Assuming such an eigenvalue exists,

$$\mathbf{X}(t) = e^{\lambda_1 t}\left[c_1\mathbf{E}_1 + c_2\mathbf{E}_2\,e^{(\lambda_2-\lambda_1)t} + \ldots + c_n\mathbf{E}_n\,e^{(\lambda_n-\lambda_1)t}\right]$$

Hence, for large t, $\mathbf{X}(t) \approx c_1\,\mathbf{E}_1\,e^{\lambda_1 t}$.

Example 2.4 For the transfer matrix **A** in Example 1.6, the eigenvalues are $\lambda = 0, -0.938$, and -1.1618. An eigenvector corresponding to $\lambda_1 = 0$ is $\mathbf{E}_1 = [0.3014, 0.4823, 0.8591]$. If material is being interchanged continuously, then $\dot{\mathbf{X}} = \mathbf{A}\,\mathbf{X}$ and, for large t,

$$\mathbf{X}(t) \approx c_1 \mathbf{E}_1 .$$

Since $\mathbf{X}(0) = [100, 0, 0]$ and since the material has no way of leaving the system, $x_1(t) + x_2(t) + x_3(t) = 100$ for all t. Hence

$$0.3014\,c_1 + 0.4823\,c_1 + 0.8591\,c_1 = 100$$

Therefore $c_1 = 60.86956$ and so $\mathbf{X}(t) \approx [18.346, 29.357, 52, 293]$. The *steady state solution* is $\lim_{t\to\infty} \mathbf{X}(t) = c_1 \mathbf{E}_1$.

2.3 REPEATED EIGENVALUES

The step-by-step procedure given on p. 31 should be adequate for solving the applied problems in the remainder of the text. Nevertheless, to be mathematically complete, we will develop more general formulas that apply whether **A** has n linearly independent eigenvalues or not. The new formulas depend on the famous *Cayley–Hamilton theorem*:

Theorem 2.4 If $p(\lambda)$ is the characteristic polynomial of the matrix **A**, then $p(\mathbf{A}) = \mathbf{0}$.

Assume that $p(\lambda) = (\lambda - \lambda_1)^{m_1} (\lambda - \lambda_2)^{m_2} \ldots (\lambda - \lambda_k)^{m_k}$ and let $p_i(\lambda)$ be the polynomial $p(\lambda)/(\lambda - \lambda_i)^{m_i}$. The rational function $1/p(\lambda)$ can be decomposed into partial fractions

$$Q_1(\lambda)/(\lambda - \lambda_1)^{m_1} + Q_2(\lambda)/(\lambda - \lambda_2)^{m_2} + \ldots + Q_k(\lambda)/(\lambda - \lambda_k)^{m_k}$$

where the degree of $Q_i < m_i$. Multiplying both sides by $p(\lambda)$ yields

$$1 = Q_1(\lambda)p_1(\lambda) + Q_2(\lambda)p_2(\lambda) + \ldots + Q_k(\lambda)p_k(\lambda)$$

It follows that

$$\mathbf{I} = Q_1(\mathbf{A})p_1(\mathbf{A}) + Q_2(\mathbf{A})p_2(\mathbf{A}) + \ldots + Q_k(\mathbf{A})p_k(\mathbf{A})$$

We will use this equation and the Cayley–Hamilton theorem to find a representation for $\mathbf{A}^m$ in terms of the eigenvalues.

$$\begin{aligned}\mathbf{A}^m &= [(\mathbf{A} - \lambda_i\mathbf{I}) + \lambda_i\mathbf{I}]^m \\ &= \sum_{j=0}^{m} \binom{m}{j} (\mathbf{A} - \lambda_i\mathbf{I})^j (\lambda_i\mathbf{I})^{m-j} \\ &= \sum_{j=0}^{m} \binom{m}{j} \lambda_i^{m-j} (\mathbf{A} - \lambda_i\mathbf{I})^j\end{aligned}$$

For $j \geqslant m_i$, $p_i(\mathbf{A})\,(\mathbf{A} - \lambda_i \mathbf{I})^j = p(\mathbf{A})\,(\mathbf{A} - \lambda_i \mathbf{I})^{j-m_i} = \mathbf{0}$. Multiplying by $p_i(\mathbf{A})$,

$$p_i(\mathbf{A})\,\mathbf{A}^m = \sum_{j=0}^{m_i-1} \binom{m}{j} \lambda_i^{m-j}\, p_i(\mathbf{A})\,(\mathbf{A} - \lambda_i \mathbf{I})^j.$$

(*Note*: When $j > m$, we let $\binom{m}{j} = 0$.) Multiplying by $Q_i(\mathbf{A})$,

$$Q_i(\mathbf{A})p_i(\mathbf{A})\mathbf{A}^m = \sum_{j=0}^{m_i-1} \binom{m}{j} \lambda_i^{m-j}\, Q_i(\mathbf{A})p_i(\mathbf{A})\,(\mathbf{A} - \lambda_i \mathbf{I})^j.$$

Summing from $i = 1$ to k,

$$\mathbf{A}^m = \sum_{i=1}^{k} Q_i(\mathbf{A})p_i(\mathbf{A})\mathbf{A}^m$$

$$= \sum_{i=1}^{k} \left(\sum_{j=0}^{m_i-1} \binom{m}{j} \lambda_i^{m-j}\, Q_i(\mathbf{A})p_i(\mathbf{A})\,(\mathbf{A} - \lambda_i \mathbf{I})^j \right)$$

Likewise, it is possible to show that

$$\mathrm{e}^{t\mathbf{A}} = \sum_{i=1}^{k} \left(\sum_{j=0}^{m_i-1} \mathrm{e}^{\lambda_i t} \frac{t^j}{j!}\, Q_i(\mathbf{A})p_i(\mathbf{A})\,(\mathbf{A} - \lambda_i \mathbf{I})^j \right)$$

Note that the matrix $\mathbf{Z}_{ij} = Q_i(\mathbf{A})p_i(\mathbf{A})\,(\mathbf{A} - \lambda_i \mathbf{I})^j$ appears in both expressions. Given an initial condition $\mathbf{X}(0) = \mathbf{X}_0$, let

$$\mathbf{E}_{ij} = \mathbf{Z}_{ij}\, \mathbf{X}_0$$

for $i = 1$ to k and $j = 0$ to $m_i - 1$. Thus we have established:

Theorem 2.5 Given that $\mathbf{X}(0) = \mathbf{X}_0$, define $\mathbf{E}_{ij}$ as above. Then

(a) the solution to $\mathbf{X}(t + \Delta t) = \mathbf{A}\,\mathbf{X}(t)$ may be written as

$$\mathbf{X}(t) = \mathbf{X}(m\,\Delta t) = \sum_{i=1}^{k} \left(\sum_{j=0}^{m_i-1} \binom{m}{j} \lambda_i^{m-j}\, \mathbf{E}_{ij} \right)$$

(b) the solution to $\dot{\mathbf{X}} = \mathbf{A}\,\mathbf{X}$ may be written as

$$\mathbf{X}(t) = \sum_{i=1}^{k} \mathrm{e}^{\lambda_i t} \left(\sum_{j=0}^{m_i-1} \frac{t^j}{j!}\, \mathbf{E}_{ij} \right)$$

These new formulas are illustrated in the final example.

Example 2.5 Solve the difference equation $\mathbf{X}(t+1) = \mathbf{A}\,\mathbf{X}(t)$ subject to $\mathbf{X}(0) = [1, 1, 1]$ where

$$\mathbf{A} = \begin{bmatrix} 1 & 1 & 0 \\ 0 & 1 & 0 \\ 1 & 0 & 0 \end{bmatrix}$$

Solution 2.5 $p(\lambda) = \lambda(\lambda - 1)^2$ and so $\lambda = 0$ and 1 are the eigenvalues. The eigenvectors corresponding to $\lambda = 1$ are all of the form $\alpha[1, 0, 1]$ and so we can find only two linearly independent eigenvectors for $\mathbf{A}$. Hence the procedure outlined on p. 31 fails. The usual partial fractions method yields

$$\frac{1}{\lambda(\lambda-1)^2} = \frac{1}{\lambda} + \frac{-\lambda+2}{(\lambda-1)^2}$$

Thus $Q_1(\lambda) = 1$ and $Q_2(\lambda) = 2 - \lambda$. Also $p_1(\lambda) = p(\lambda)/\lambda = (\lambda - 1)^2$ and $p_2(\lambda) = p(\lambda)/(\lambda - 1)^2 = \lambda$. Hence

$$\begin{aligned} \mathbf{E}_{10} &= Q_1(\mathbf{A})p_1(\mathbf{A})\,(\mathbf{A} - 0\mathbf{I})^0\,\mathbf{X}(0) \\ &= (\mathbf{A} - \mathbf{I})^2\,\mathbf{X}(0) = [0, 0, 1] \\ \mathbf{E}_{20} &= Q_2(\mathbf{A})p_2(\mathbf{A})\,(\mathbf{A} - \mathbf{I})^0\,\mathbf{X}(0) \\ &= (2\mathbf{I} - \mathbf{A})\mathbf{A}\,\mathbf{X}(0) = [1, 1, 0] \end{aligned}$$

and

$$\begin{aligned} \mathbf{E}_{21} &= Q_2(\mathbf{A})p_2(\mathbf{A})\,(\mathbf{A} - \mathbf{I})\mathbf{X}(0) \\ &= (2\mathbf{I} - \mathbf{A})\mathbf{A}(\mathbf{A} - \mathbf{I})\mathbf{X}(0) = [1, 0, 1] \end{aligned}$$

It follows from Theorem 2.5(a) that

$$\begin{aligned} \mathbf{X}(t) = \mathbf{X}(n\,\Delta t) &= 0\,\mathbf{E}_{10} + \mathbf{E}_{20} + n\,\mathbf{E}_{21} \\ &= [1, 1, 0] + [n, 0, n] \\ &= [n+1, 1, n] \end{aligned}$$

EXERCISES

Part A

1 Find the general solution to the differential equation $\dot{\mathbf{X}} = \mathbf{A}\,\mathbf{X}$ where

$$\mathbf{A} = \begin{bmatrix} 2 & 3 \\ 2 & 1 \end{bmatrix}$$

2 Find the general solution to the difference equation $\mathbf{X}(t+1) = \mathbf{A}\,\mathbf{X}(t)$ where

$$\mathbf{A} = \begin{bmatrix} 1 & -3 \\ -2 & 2 \end{bmatrix}$$

3 Solve the differential equation $\dot{\mathbf{X}} = \mathbf{B}\,\mathbf{X}$ subject to the initial condition $\mathbf{X}(0) = [10, 5, 0]$ if

$$\mathbf{B} = \begin{bmatrix} 0 & 6 & 0 \\ 1 & 0 & 1 \\ 1 & 1 & 0 \end{bmatrix}$$

4 Solve the difference equation $\mathbf{X}(t+1) = \mathbf{C}\,\mathbf{X}(t)$ subject to the initial condition $\mathbf{X}(0) = [0, 0, 100]$ if $\mathbf{C}$ is the matrix

$$\mathbf{C} = \begin{bmatrix} 2 & 2 & 1 \\ 1 & 3 & 1 \\ 1 & 2 & 2 \end{bmatrix}$$

5 Solve the differential equation $\dot{\mathbf{X}} = \mathbf{A}\,\mathbf{X}$ with $\mathbf{X}(0) = [10, 0, 0]$ if $\mathbf{A}$ is the matrix

$$\mathbf{A} = \begin{bmatrix} -1 & 0 & -1 \\ 0 & -1 & 0 \\ -1 & 0 & -1 \end{bmatrix}$$

What is $\lim_{t \to \infty} \mathbf{X}(t)$?

6 If $\mathbf{A}$ is the matrix $\begin{bmatrix} -1 & 1 & 0 \\ 1 & 2 & 1 \\ 0 & 3 & -1 \end{bmatrix}$, describe the solution to $\dot{\mathbf{X}} = \mathbf{A}\,\mathbf{X}$ for large t.

7 Find the general solution to $\mathbf{X}(t+1) = \mathbf{F}\,\mathbf{X}(t)$ if $\mathbf{F}$ is the matrix

$$\mathbf{F} = \begin{bmatrix} 3 & 2 & 2 & -4 \\ 2 & 3 & 2 & -1 \\ 1 & 1 & 2 & -1 \\ 2 & 2 & 2 & -1 \end{bmatrix}$$

8 Find the general solution to $\dot{\mathbf{X}} = \mathbf{A}\,\mathbf{X}$ with $\mathbf{A} = \begin{bmatrix} 4 & 1 \\ -9 & -2 \end{bmatrix}$.

9 Find the general solution to the difference equation $\mathbf{X}(t+1) = \mathbf{A}\,\mathbf{X}(t)$ where

$$\mathbf{A} = \begin{bmatrix} 3 & -1 \\ 9 & -3 \end{bmatrix}$$

10 Find the solution to $\dot{\mathbf{X}} = \mathbf{A}\,\mathbf{X}$ subject to $\mathbf{X}(0) = [100, 0, 0]$ if $\mathbf{A}$ is the matrix

$$\mathbf{A} = \begin{bmatrix} 5 & -4 & 0 \\ 1 & 0 & 2 \\ 0 & 2 & 5 \end{bmatrix}$$

11 Let $\mathbf{A}$ be the ecomatrix

$$\begin{bmatrix} -0.1 & 0.1 & 0 \\ 0.1 & -0.1 & 0.2 \\ 0 & 0 & -0.2 \end{bmatrix}$$

(a) If $\mathbf{X}(0) = [0, 0, 10]$, find the solution to $\dot{\mathbf{X}} = \mathbf{A}\,\mathbf{X}$.

(b) What is $\lim_{t\to\infty} \mathbf{X}(t)$?

(c) When are the contents of compartment 2 a maximum?

12 The general 2×2 ecomatrix takes the form

$$\begin{bmatrix} -a & b \\ a & -b \end{bmatrix}.$$

(a) Solve the system $\dot{\mathbf{X}} = \mathbf{A}\,\mathbf{X}$ and compute $\lim_{t\to\infty} \mathbf{X}(t)$ given that $\mathbf{X}(0) = [1, 0]$.

(b) If $\mathbf{X}(0) = [c, d]$, with $c + d = 1$, what is $\lim_{t\to\infty} \mathbf{X}(t)$?

Part B Eigenvalues of an ecomatrix

If $\mathbf{A}$ is an irreducible ecomatrix, then it is known that the non-zero eigenvalues satisfy the property that Re $\lambda < 0$, where Re denotes the real part of λ. (See [1], pages 30, 54, and 261.)

13 If $\mathbf{A}$ is an irreducible ecomatrix and 0 is an eigenvalue of multiplicity one, prove that if $\mathbf{X}(t)$ is a solution to $\dot{\mathbf{X}} = \mathbf{A}\,\mathbf{X}$, then $\hat{\mathbf{X}} = \lim_{t\to\infty} \mathbf{X}(t)$ exists.

14 If $\mathbf{A}$ is a 3×3 ecomatrix, find conditions on the transfer coefficients that ensure that all eigenvalues are real.

REFERENCES

[1] R. S. Varga, *Matrix Iterative Analysis*, Prentice-Hall (1960).

CHAPTER 3

Computer Implementation

3.1 THE NECESSARY SUBROUTINES

In Chapter 2, we were able to obtain eigenvalue solutions easily only because the matrices were of small order and had a characteristic polynomial $p(\lambda)$ that happened to factor nicely. In real applications, the matrices are often of large order. For example, in Chapter 4, we will need to manipulate an 18×18 matrix to make predictions about the United States population. In addition, the matrix entries a_{ij} are often estimated from statistical data. Therefore, even if we are able to compute $p(\lambda)$, it is highly unlikely that we can find the zeros using algebra. Here we need algorithms from numerical analysis and the computer to assist us in performing the steps listed on page 31 or in using the formulas $\mathbf{X}(m\ \Delta t) = \mathbf{A}^m\ \mathbf{X}(0)$ and $\mathbf{X}(t) = e^{\mathbf{A}t}\ \mathbf{X}(0)$ from Chapter 1.

Your computer's software library should have the necessary scientific subroutines. If you will be programming in FORTRAN, an excellent collection of programs is available in the IMSL (International Mathematical and Statistical Library) subroutine library. If you are using a form of BASIC, you should check and see if the matrix MAT commands are present. They can make the programming much easier. Now let us examine exactly what is needed at each step in the solution process.

Steps 1–3: The Eigenvalue–Eigenvector Analysis The IMSL subroutine called EIGRF is a sophisticated numerical method for finding real and complex eigenvalues. The input is a real $n \times n$ matrix $\mathbf{A}$. The output is a complex vector $\mathbf{W} = [\lambda_1, \lambda_2, \ldots, \lambda_n]$ containing eigenvalues of $\mathbf{A}$ and a complex matrix $\mathbf{Z}$

whose ith column is a complex eigenvector corresponding to λ_i. We will also develop an elementary eigenvalue–eigenvector technique known as the *power method* later in this chapter. Although there are many instances where this method fails, the power method is easy to implement on the computer and should be adequate for most problems.

Often there are a large number of zeros in the matrix $\mathbf{A}$ with the result that the characteristic polynomial $p(\lambda)$ can be explicitly found. To complete step 2, we need a program that will find all zeros of a real polynomial. In the IMSL library, such a program is ZRPOLY. Another program, commonly called BAIRSTOW, is based on Bairstow's iteration method, a generalization of the Newton's method algorithm of calculus ([1], p. 76–83).

To find the corresponding eigenvectors (i.e. solve $(\mathbf{A} - \lambda\mathbf{I})\mathbf{X} = \mathbf{0}$), we need a program for solving linear systems of the form $\mathbf{AX} = \mathbf{B}$. In the IMSL library, the subroutine is LEQT1C. The input is a real or complex $n \times n$ matrix $\mathbf{A}$ and a real or complex $n \times m$ matrix $\mathbf{B}$. The output is an $n \times m$ solution matrix $\mathbf{X}$. Of course we will use the program with $\mathbf{A} - \lambda\mathbf{I}$ replacing $\mathbf{A}$ and with the $n \times 1$ zero vector as $\mathbf{B}$. If this program is not available to you, look for a subroutine that solves linear systems by Gaussian elimination. Such a routine should prove adequate.

Steps 4–5: Computing $\mathbf{Z}^{-1}$ *and* $\mathbf{Z}^{-1}\mathbf{X}(0)$ To solve the system $\mathbf{X}(0) = \mathbf{ZC}$ for $\mathbf{C}$, all we need to do is apply the algorithm found for solving $\mathbf{AX} = \mathbf{B}$ with $\mathbf{A} = \mathbf{Z}$ and $\mathbf{B} = \mathbf{X}(0)$. Alternately, we can find $\mathbf{Z}^{-1}$ and then compute $\mathbf{Z}^{-1}\mathbf{X}(0)$. This is especially easy if you are working in a BASIC with MAT commands:

```
MAT U = INV(Z)
MAT C = U*X
MAT PRINT C
```

Shown in Table 3.1 are the MAT commands in extended basic.

3.2 SPECTRAL DECOMPOSITIONS AND THE POWER METHOD

An elementary method for finding eigenvalues and eigenvectors can be based on the *spectral decomposition theorem* for matrices:

Theorem 3.1 Let $\mathbf{A}$ be an $n \times n$ matrix with eigenvalues $\lambda_1, \ldots, \lambda_n$ and linearly independent eigenvectors $\mathbf{E}_1, \ldots, \mathbf{E}_n$. Then there are $n \times n$ matrices $\mathbf{Z}_1, \ldots, \mathbf{Z}_n$ with

$$\mathbf{A} = \lambda_1\mathbf{Z}_1 + \lambda_2\mathbf{Z}_2 + \ldots + \lambda_n\mathbf{Z}_n$$

and

$$\mathbf{Z}_i\mathbf{Z}_j = \begin{cases} 0, & \text{for } i \neq j \\ \mathbf{Z}_i, & \text{for } i = j \end{cases}$$

$\mathbf{Z}_i$ is called the ith *spectral component* for $\mathbf{A}$.

Table 3.1

Command	Purpose
MAT B = A	Forms a second matrix **B** equal to **A**. No dimension statement for **B** is needed.
MAT A = ZER[N, M]	Forms the $N \times M$ zero matrix.
MAT A = CON[N, M]	Forms an $N \times M$ matrix all of whose entries equal 1.
MAT A = IDN[N, N]	Forms the $N \times N$ identity matrix.
MAT READ A	Reads matrix entries from data list row by row.
MAT INPUT A	Reads matrix entries entered row by row from the keyboard.
MAT PRINT A	Prints the matrix **A**.
MAT C = A ± B	Adds or subtracts matrices **A** and **B**. It is permissible to write MAT A = A + B.
MAT C = A*B	Multiplies compatible matrices **A** and **B**. It is permissible to write MAT B = A*B.
MAT C = (A)*B	Multiplies the matrix **B** by the *scalar* A.
MAT B = INV(A)	Computes the inverse of the matrix **A**. It is permissible to write MAT A = INV(A).
D = DET(A)	Computes the determinant of an $N \times N$ matrix whose inverse has just been taken.
MAT B = TRN(A)	Forms the transpose of an $N \times M$ matrix **A**. No dimension statement for **B** is needed.

We have seen in Theorem 2.3 that the hypotheses of the theorem are true when all n eigenvalues are distinct.

Proof of Theorem 3.1 Let $\mathbf{Z}$ be the matrix $\left[\mathbf{E}_1 \mid \mathbf{E}_2 \mid \ldots \mid \mathbf{E}_n\right]$ of eigenvectors. It follows that

$$\mathbf{Z}^{-1}\,\mathbf{A}\,\mathbf{Z} = \begin{bmatrix} \lambda_1 & & & \mathbf{0} \\ & \lambda_2 & & \\ & & \ddots & \\ \mathbf{0} & & & \lambda_n \end{bmatrix}$$

To see why this is valid, let $\mathbf{X}_i = [0, 0, \ldots, 0, 1, 0, \ldots, 0]$, where 1 occurs in the ith position. Then $\mathbf{Z}\mathbf{X}_i = \mathbf{E}_i$ and so

$$(\mathbf{Z}^{-1}\,\mathbf{A}\,\mathbf{Z})\mathbf{X}_i = \mathbf{Z}^{-1}\,\mathbf{A}\,\mathbf{E}_i = \mathbf{Z}^{-1}(\lambda_i\mathbf{E}_i) = \lambda_i(\mathbf{Z}^{-1}\,\mathbf{E}_i) = \lambda_i\,\mathbf{X}_i$$

Thus the ith column of $\mathbf{Z}^{-1}\,\mathbf{A}\,\mathbf{Z}$ is formed from $\lambda_i\mathbf{X}_i = [0, \ldots, \lambda_i, \ldots, 0]$.

Next let $\mathbf{I}_k$ be the $n \times n$ matrix with all 0s except for a 1 in the kk position.

Then

$$\mathbf{I} = \mathbf{I}_1 + \mathbf{I}_2 + \ldots + \mathbf{I}_n$$

and

$$\mathbf{I}_i \mathbf{I}_j = \begin{cases} 0, & \text{for } i \neq j \\ \mathbf{I}_i, & \text{for } i = j \end{cases}.$$

It follows that

$$\mathbf{Z}^{-1}\,\mathbf{A}\,\mathbf{Z} = \sum_{k=1}^{n} \lambda_k\,\mathbf{I}_k \quad \text{or} \quad \mathbf{A} = \sum_{k=1}^{n} \lambda_k\,(\mathbf{Z}\,\mathbf{I}_k\,\mathbf{Z}^{-1})$$

Letting $\mathbf{Z}_k = \mathbf{Z}\,\mathbf{I}_k\,\mathbf{Z}^{-1}$, then $\mathbf{Z}_i\,\mathbf{Z}_j = (\mathbf{Z}\mathbf{I}_i\,\mathbf{Z}^{-1})\,(\mathbf{Z}\mathbf{I}_j\,\mathbf{Z}^{-1}) = \mathbf{Z}(\mathbf{I}_i\mathbf{I}_j)\mathbf{Z}^{-1}$.

It follows that $\mathbf{A} = \sum_{k=1}^{n} \lambda_k\,\mathbf{Z}_k$ and $\mathbf{Z}_i\,\mathbf{Z}_j = \begin{cases} 0, & \text{for } i \neq j \\ \mathbf{Z}_i, & \text{for } i = j \end{cases}.$

Corollary 3.1.1 Given the spectral decomposition for $\mathbf{A}$,

$$\mathbf{A}^m = \lambda_1^m\,\mathbf{Z}_1 + \lambda_2^m\,\mathbf{Z}_2 + \ldots + \lambda_n^m\,\mathbf{Z}_n.$$

Proof The result follows by induction and from the fact that for any choice of a_i and b_j,

$$\left(\sum_{i=1}^{n} a_i\,\mathbf{Z}_i\right)\left(\sum_{j=1}^{n} b_j\,\mathbf{Z}_j\right) = \sum_{i=1}^{n}\left(\sum_{j=1}^{n} a_i\,b_j\,\mathbf{Z}_i\,\mathbf{Z}_j\right)$$

$$= \sum_{i=1}^{n} a_i\,b_i\,\mathbf{Z}_i$$

Corollary 3.1.2 The columns of the spectral component $\mathbf{Z}_i$ are eigenvectors corresponding to λ_i.

Proof $\mathbf{A}\mathbf{Z}_i = \left(\sum_{k=1}^{n} \lambda_k\,\mathbf{Z}_k\right)\mathbf{Z}_i = \sum_{k=1}^{n} \lambda_k(\mathbf{Z}_k\,\mathbf{Z}_i) = \lambda_i\,\mathbf{Z}_i^2 = \lambda_i\,\mathbf{Z}_i$. Hence

if $\mathbf{Z}_i = \left[\mathbf{C}_1 \vdots \mathbf{C}_2 \vdots \ldots \vdots \mathbf{C}_n\right]$,

$$\mathbf{A}\,\mathbf{Z}_i = \left[\mathbf{A}\mathbf{C}_1 \vdots \mathbf{A}\mathbf{C}_2 \vdots \ldots \vdots \mathbf{A}\mathbf{C}_n\right] = \lambda_i\,\mathbf{Z}_i = \left[\lambda_i\,\mathbf{C}_1 \vdots \ldots \vdots \lambda_i\,\mathbf{C}_n\right]$$

It follows that $\mathbf{A}\mathbf{C}_k = \lambda_i\,\mathbf{C}_k$ for any column $\mathbf{C}_k$ of $\mathbf{Z}_i$.

Suppose that we have ordered and named the eigenvalues to that $|\lambda_1| \geqslant |\lambda_2| \geqslant \ldots \geqslant |\lambda_n|$. If $|\lambda_1| > |\lambda_i|$ for $i = 2$ to n, we say that λ_1 is the *dominant eigenvalue* for $\mathbf{A}$.

Corollary 3.1.3 If $\mathbf{A}$ has a dominant eigenvalue λ_1, then

$$\lim_{m\to\infty}\left(\frac{1}{\lambda_1}\mathbf{A}\right)^m = \mathbf{Z}_1.$$

Proof By Corollary 3.1.1, we may write

$$\left(\frac{1}{\lambda_1}\mathbf{A}\right)^m = \mathbf{Z}_1 + \left(\frac{\lambda_2}{\lambda_1}\right)^m \mathbf{Z}_2 + \ldots + \left(\frac{\lambda_n}{\lambda_1}\right)^m \mathbf{Z}_n$$

Since $|\lambda_i/\lambda_1| < 1$, $\lim_{m\to\infty} (\lambda_i/\lambda_1)^m = 0$, for $i = 2$ to n, and the result follows.

Corollary 3.1.3 can be used to justify the power method for finding eigenvalues and eigenvectors. If a dominant eigenvalue exists, then, for large m, $\mathbf{A}^m \approx \lambda_1^m \mathbf{Z}_1$ and $\mathbf{A}^{m+1} \approx \lambda_1^{m+1} \mathbf{Z}_1$, and so $\mathbf{A}^{m+1} \approx \lambda_1 \mathbf{A}^m$. If we let $\mathbf{A}^m = [a_{ij}^{(m)}]$, then we can make two conclusions:

1. λ_1 is approximated by $a_{ij}^{(m+1)}/a_{ij}^{(m)}$ when $a_{ij}^{(m)} \neq 0$.
2. $\mathbf{Z}_1$ is approximated by $\left(\frac{1}{\lambda_1}\mathbf{A}\right)^m$.

Having found λ_1 and $\mathbf{Z}_1$ to a desired number of decimal places, we can let $\mathbf{B} = \mathbf{A} - \lambda_1\mathbf{Z}_1 = \lambda_2\mathbf{Z}_2 + \ldots + \lambda_n\mathbf{Z}_n$ and (if possible) repeat the process.

In the case where $\lambda_2 = -\lambda_1$, the power method can be adapted to obtain λ_1 and an eigenvector. If λ_1 is complex, however, $\overline{\lambda}_1$ will also be an eigenvalue and the sequence $\left(\frac{1}{\lambda_1}\mathbf{A}\right)^m$ does not converge. The power method works best when *all eigenvalues are real and distinct*. This is illustrated in our final example.

Example 3.1 Let $\mathbf{A}$ be the matrix

$$\begin{bmatrix} 0.2 & 3 & 1 \\ 3 & -0.5 & 0 \\ 1 & 0 & -2 \end{bmatrix}$$

Computing high powers of $\mathbf{A}$ and forming ratios $a_{ij}^{(m+1)}/a_{ij}^{(m)}$, we obtain (to four correct decimal places) $\lambda_1 = -3.48641$ and

$$\mathbf{Z}_1 = \begin{bmatrix} 0.4062 & -0.4081 & -0.2733 \\ -0.4081 & 0.4099 & 0.2745 \\ -0.2733 & 0.2745 & 0.1839 \end{bmatrix}$$

We then let $\mathbf{B} = \mathbf{A} - \lambda_1\mathbf{Z}_1$ and repeat the process to obtain $\lambda_2 = 2.98393$ and

$$Z_2 = \begin{bmatrix} 0.5612 & 0.4833 & 0.1126 \\ 0.4833 & 0.4162 & 0.0970 \\ 0.1126 & 0.0970 & 0.0226 \end{bmatrix}$$

Letting $\mathbf{C} = \mathbf{B} - \lambda_2 \mathbf{Z}_2$ and raising $\mathbf{C}$ to high powers yields $\lambda_3 = -1.7975$ and

$$Z_3 = \begin{bmatrix} 0.0325 & -0.0752 & 0.1607 \\ -0.0752 & 0.1739 & -0.3715 \\ 0.1607 & -0.3715 & 0.79355 \end{bmatrix}$$

You can check that $\lambda_1 \mathbf{Z}_1 + \lambda_2 \mathbf{Z}_2 + \lambda_3 \mathbf{Z}_3 = \mathbf{A}$. Since the columns of the spectral components are eigenvectors,

$$Z = \begin{bmatrix} 0.4062 & 0.5612 & 0.0325 \\ -0.4081 & 0.4833 & -0.0752 \\ -0.2733 & 0.1126 & 0.1607 \end{bmatrix}$$

is a matrix of eigenvectors.

EXERCISES

Programming exercises

The programs requested in Exercises 1 and 2 should prove useful in doing many of the exercises that follow.

1 Write a program that implements the power method for finding eigenvalues. Output the eigenvalues and spectral components.

2 Write a program which forms the matrix $\mathbf{Z}$ whose columns are the eigenvectors of $\mathbf{A}$, finds the inverse of $\mathbf{Z}$, and computes

$$\mathbf{C} = \mathbf{Z}^{-1}\ \mathbf{X}(0).$$

Part A

Find the *spectral decomposition* of each of the following matrices.

3 $\begin{bmatrix} 2 & 3 \\ 2 & 1 \end{bmatrix}$ **4** $\begin{bmatrix} 1 & 4 \\ 1 & 1 \end{bmatrix}$

5 $\begin{bmatrix} 1 & -3 \\ -2 & 2 \end{bmatrix}$ **6** $\begin{bmatrix} 0 & 6 & 0 \\ 1 & 0 & 1 \\ 1 & 1 & 0 \end{bmatrix}$

7 $\begin{bmatrix} 1 & 2 & 1 \\ 6 & -1 & 0 \\ -1 & -2 & -1 \end{bmatrix}$ 8 $\begin{bmatrix} 0.2 & 0.5 & 0 \\ 0.3 & 0.1 & 0.4 \\ 0.5 & 0.4 & 0.6 \end{bmatrix}$

9 Assuming continuous transfers, find the eigenvalue solution to the compartmental model in Exercise 4 of Chapter 1. What is $\lim_{t\to\infty} \mathbf{X}(t)$?

10 Find the eigenvalue solution to the compartmental model in Exercise 5, Chapter 1. What is $\lim_{t\to\infty} \mathbf{X}(t)$?

11 Find the eigenvalue solution to the compartmental model in Exercise 6, Chapter 1 assuming discrete transfers with $\Delta t = 1$ year. What is $\lim_{t\to\infty} \mathbf{X}(t)$?

12 The following system of differential equations is taken from a compartmental model by Patten and Witkamp [2]:

$$\dot{\mathbf{X}}(t) = \begin{bmatrix} -0.199 & 0 & 0.500 & 0.250 & 0 \\ 0.033 & 0 & 0 & 0.160 & 0 \\ 0.120 & 0 & -0.750 & 0.250 & 0 \\ 0.045 & 0 & 0.250 & -0.660 & 0 \\ 0.001 & 0 & 0 & 0 & 0 \end{bmatrix} \mathbf{X}(t)$$

subject to $\mathbf{X}(0) = [1, 0, 0, 0, 0]$.

(a) Show that the non-zero eigenvalues of $\mathbf{A}$ are the eigenvalues of the matrix

$$\begin{bmatrix} -0.199 & 0.500 & 0.250 \\ 0.120 & -0.750 & 0.250 \\ 0.045 & 0.250 & -0.660 \end{bmatrix}$$

(*Hint*: Expand det $(\lambda\mathbf{I} - \mathbf{A})$ along the last column and then the second column of the minor.)

(b) Find the eigenvalues of the matrix in (a).
(c) Find two linearly independent eigenvectors corresponding to $\lambda = 0$.
(d) Compute $\lim_{t\to\infty} \mathbf{X}(t)$.

Part B

13 If $\lambda_2 = -\lambda_1$ in the spectral decomposition for $\mathbf{A}$ and if $|\lambda_1| > |\lambda_k|$ for $k = 3, \ldots, n$, then prove that $\mathbf{A}^m \approx \lambda_1^m [\mathbf{Z}_1 + (-1)^m \mathbf{Z}_2]$ for large m. Conclude that $\lambda_1^2 \approx a_{ij}^{(m+2)}/a_{ij}^{(m)}$, when $a_{ij}^{(m)} \neq 0$.

14 If $\mathbf{A}$ is invertible, find the spectral decomposition of $\mathbf{A}^{-1}$ in terms of the spectral decomposition of $\mathbf{A}$.

15 If $\mathbf{A} = \lambda_1 \mathbf{Z}_1 + \ldots + \lambda_n \mathbf{Z}_n$ is the spectral decomposition for $\mathbf{A}$, show that $e^{t\mathbf{A}} = e^{\lambda_1 t}\, \mathbf{Z}_1 + \ldots + e^{\lambda_n t}\, \mathbf{Z}_n$.

REFERENCES

[1] Melvin J. Maron, *Numerical Analysis: A Practical Approach*, Macmillan Publishing Company (1982).

[2] B. C. Patten and M. Witkamp, 'Systems Analysis of ^{134}Cesium Kinetics in Terrestrial Microcosms', *Ecology* (1967), **48**, 813–824.

CHAPTER 4

The Leslie Matrix Model

4.1 MODEL ASSUMPTIONS AND EQUATIONS

In 1945, the English biologist P. H. Leslie introduced a model that takes into account the *age structure* of a population [7]. Earlier versions of the model were presented by E. Lewis ([8], 1941) and H. Bernadelli ([1], 1942), but it was Leslie who developed the model in detail and popularized its use. This basic model is widely used in population dynamics and especially human demography. In the first version of the model, *only females are considered*.

Assume that the population is divided into $n + 1$ age categories, each of length k years. By the age of an animal, we mean *age last birthday*. Thus the age categories are

$$C_0 = [0, k),\ C_1 = [k, 2k),\ C_2 = [2k, 3k), \ldots, C_n = [kn, k(n+1))$$

Let $x_i(t)$ = number in age class C_i at time t, and define†

$$\mathbf{X}(t) = [x_0(t), x_1(t), \ldots, x_n(t)].$$

The vector $\mathbf{X}(t)$ specifies the age structure of the population. For human populations, k is usually 5 and $n = 17$. Thus the age classes are 0–4 years, 5–9 years, . . . , 85–89 years. For large whales k is usually 2. On the other hand, for rodents, k may be as small as 1/12 (or 1 month).

† We will regard $\mathbf{X}(t)$ as either a row or column vector and therefore will not distinguish between $\mathbf{X}(t)$ and the transpose of $\mathbf{X}(t)$.

Given $\mathbf{X}(0)$, we wish to predict the population structure at times $\Delta t = k$, $2\Delta t = 2k, \ldots$ Thus the time unit is $\Delta t = k$ years. It is convenient, however, to call one unit of time k years and write $\mathbf{X}(t+1)$ rather than $\mathbf{X}(t + \Delta t)$. We will now show that $\mathbf{X}(t+1) = \mathbf{P}\,\mathbf{X}(t)$ for an $n \times n$ matrix $\mathbf{P}$ whose entries are given in terms of *survival* and *fecundity rates*. Define

S_0 = probability of surviving from age class C_0 to age class C_1 k years later

S_1 = probability of surviving from age class C_1 to age class C_2 k years later

.
.
.

S_{n-1} = probability of surviving from age class C_{n-1} to the final age class C_n k years later.

Note that each S_k is assumed to be constant over time and independent of the total number in the population. Thus, the model does not take crowding into account. In addition, the model does not keep track of those that live more than $k(n+1)$ years. We also assume that $S_k > 0$ for each k. Otherwise, we could reduce the number of age classes.

Let F_k be the average number of female offspring, born to an individual in age class C_k, that survive to the next census. As we will illustrate later in the chapter, F_k is determined not only by the average number of female births per mother per time period, but also by infant survival rates. F_k is again assumed to be constant and therefore density and time independent. As illustrated in Fig. 4.1, note that

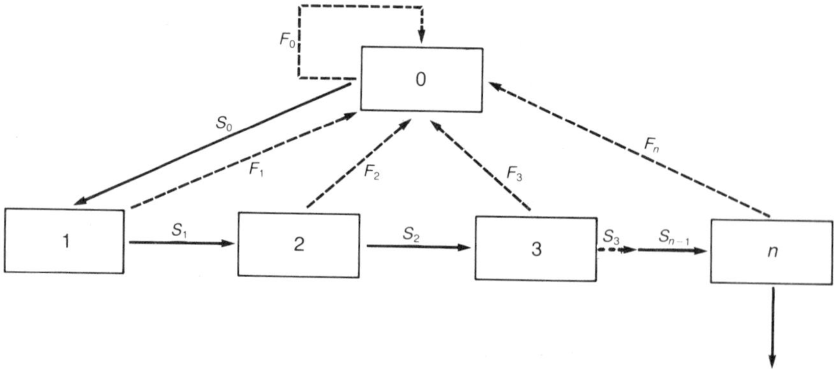

Fig. 4.1: The Leslie Matrix Model

$$x_0(t+1) = F_0 x_0(t) + F_1 x_1(t) + \ldots + F_n x_n(t)$$

$$x_1(t+1) = S_0 x_0(t)$$

$$x_2(t+1) = \qquad S_1 x_1(t)$$

$$\vdots$$

$$x_n(t+1) = \qquad\qquad S_{n-1}\, x_{n-1}(t)$$

In matrix form, $\mathbf{X}(t+1) = \mathbf{P}\,\mathbf{X}(t)$ where

$$\mathbf{P} = \begin{bmatrix} F_0 & F_1 & F_2 & \cdots & F_n \\ S_0 & 0 & 0 & \cdots & 0 \\ 0 & S_1 & 0 & \cdots & 0 \\ \vdots & \vdots & \vdots & & \vdots \\ 0 & 0 & 0 & S_{n-1} & 0 \end{bmatrix}$$

How realistic are the assumptions? In the case of human populations, the survival rates, especially amongst the elderly, will change with every new medical advance. In addition, new social attitudes towards marriage and the family can have profound effects on the fecundity rates F_k. It is important to realize that we are assuming the *status quo* when making our population predictions.

Seasonal differences in survival rates and the density dependence of the survival and fecundity rates, although they do exist, are not that significant in the case of human populations. For the African elephant, however, both F_k and S_k are very much a function of the total $T(t) = x_0(t) + x_1(t) + \ldots + x_n(t)$ in the population. If D is the density (in elephants/square mile), the age of maturity is $10 + D$ years and the probability of an infant surviving the first year is $1 - 0.03\,D$. The fecundity coefficients are density dependent and are given by

$$F_k = \begin{cases} 0, & \text{for } D > 33 \\ \dfrac{1 - 0.03\,D}{4\,D^{1/3}} & \end{cases}$$

for age $k \geqslant 10 + D$. Our basic matrix model does not apply and must be altered (see [3]).

Example 4.1 (Bernadelli's beetles) In his 1941 paper [1], H. Bernadelli gave the following Leslie matrix $\mathbf{P}$ for a fictitious beetle population:

$$\mathbf{P} = \begin{bmatrix} 0 & 0 & 6 \\ 1/2 & 0 & 0 \\ 0 & 1/3 & 0 \end{bmatrix}$$

If $\mathbf{X}(0) = [0, 0, 18]$ and $k = 1$ month, then the population projections are shown in Table 4.1. Note that the population cycles and that only one age class is represented at each census. Such a population is called *unstable*. The characteristic polynomial of $\mathbf{P}$ is $p(\lambda) = \lambda^3 - 1$ and so $\mathbf{P}$ has eigenvalues 1 and $-1/2 \pm (\sqrt{3}/2)i$. All eigenvalues have absolute value 1. We will see shortly that this lack of a dominant eigenvalue results in cyclical behavior.

Table 4.1

Month	Age class 1	Age class 2	Age class 3
0	0	0	18
1	108	0	0
2	0	54	0
3	0	0	18
4	108	0	0
5	0	54	0
6	0	0	18

Adapted from M. R. Cullen, *Mathematics for the Biosciences*, PWS Publishers, 1983.

4.2 EIGENVALUE SOLUTIONS

If **P** has eigenvalues $\lambda_0, \lambda_1, \ldots, \lambda_n$ and corresponding linearly independent eigenvectors $\mathbf{E}_0, \mathbf{E}_1, \ldots, \mathbf{E}_n$, then solutions to $\mathbf{X}(t+1) = \mathbf{P}\,\mathbf{X}(t)$ are of the form

$$\mathbf{X}(t) = c_0\mathbf{E}_0(\lambda_0)^t + c_1\mathbf{E}_1(\lambda_1)^t + \ldots + c_n\mathbf{E}_n(\lambda_n)^t$$

If λ_0 is a dominant eigenvalue for **P**, we can write $\mathbf{X}(t)$ as

$$\mathbf{X}(t) = \lambda_0^t \left[c_0\mathbf{E}_0 + \sum_{k=1}^{n} c_i\mathbf{E}_i(\lambda_k/\lambda_0)^t \right]$$

and conclude that for large t, $\mathbf{E}(t) \approx c_0\mathbf{E}_0(\lambda_0)^t$. If we let

$$c_0\mathbf{E}_0 = [y_0, y_1, \ldots, y_n]$$

then $x_k(t) \approx y_k(\lambda_0)^t$. Hence, *if a dominant eigenvalue exists, we may conclude that each age class eventually will grow exponentially at the rate of λ_0 per time period*.

Let $y = y_0 + y_1 + \ldots + y_n$ and form the eigenvector

$$\mathbf{S} = \frac{1}{y}(c_0\mathbf{E}_0) = [y_0/y, y_1/y, \ldots, y_n/y]$$

The vector **S** is called the *stable age distribution* of the population. This designation is appropriate since, for large t, y_k/y is the proportion of the population in age class C_k. The approach to **S** depends on exactly how much larger λ_0 is than $|\lambda_1|, \ldots, |\lambda_n|$. This is illustrated in our next example.

Example 4.2 A population is divided into three age classes and Leslie matrix **P** is given as

$$\mathbf{P} = \begin{bmatrix} 0 & 9 & 12 \\ 1/3 & 0 & 0 \\ 0 & 1/2 & 0 \end{bmatrix}$$

The characteristic polynomial of **P** is easily seen to be $p(\lambda) = \lambda^3 - 3\lambda - 2 = (\lambda - 2)(\lambda + 1)^2$. Thus **P** has eigenvalues 2 and -1. To find the stable age distribution **S**, we must first find an eigenvector corresponding to $\lambda = 2$. $\mathbf{PX} = 2\mathbf{X}$ implies

$$9x_1 + 12x_2 = 2x_0$$
$$x_0/3 = 2x_1$$
$$x_1/2 = 2x_2$$

If we let $x_2 = 1$, then $x_1 = 4$, and $x_0 = 6x_1 = 24$. Hence $\mathbf{E}_0 = [24, 4, 1]$ and so

$$\mathbf{S} = [24/29, 4/29, 1/29]$$
$$= [0.828, 0.138, 0.034]$$

Since the second eigenvalue is -1, the stable age distribution should be reached quickly. Table 4.2 shows the age distribution after each time period when $\mathbf{X}(0) = [10, 8, 5]$.

Table 4.2

	Age distribution		
Time t	Class 1	Class 2	Class 3
0	0.435	0.348	0.217
1	0.947	0.024	0.029
2	0.631	0.356	0.013
3	0.897	0.056	0.047
4	0.767	0.213	0.020
5	0.856	0.101	0.042
6	0.808	0.163	0.029
7	0.838	0.125	0.038
8	0.822	0.146	0.033
9	0.831	0.134	0.036
10	0.826	0.140	0.034

In most cases, we will need the assistance of a computer to find λ_0, λ_1, and **S**. We will develop theorems that will enable us to find the characteristic polynomial and all eigenvectors of a Leslie matrix **P** quite easily. Before doing so, let us turn to a more fundamental problem.

4.3 ESTIMATING MODEL PARAMETERS

Exactly how does one calculate the fecundity constants F_k? In the case of continually breeding populations such as human populations the situation is complicated. Since we are counting only those babies that are alive after five years, we must take into account not only reproductive rates but also infant mortality and the mortality of young children aged 1 to 4. For more information, see [5].

There is one commonly occurring case that is easy to analyze. Most species can reproduce during a relatively short period of time. As an example, the blue whale can conceive during the time period between roughly 20 June and 20 August in the warm waters near the equator. Since the gestation period lasts a year and since young whales are nursed for seven months, conception is possible only once every two years during a two-month period.

Let M_k = average number of female offspring per female in age class C_k born between t and $t + \Delta t$.

Case I We census the population just before reproduction.

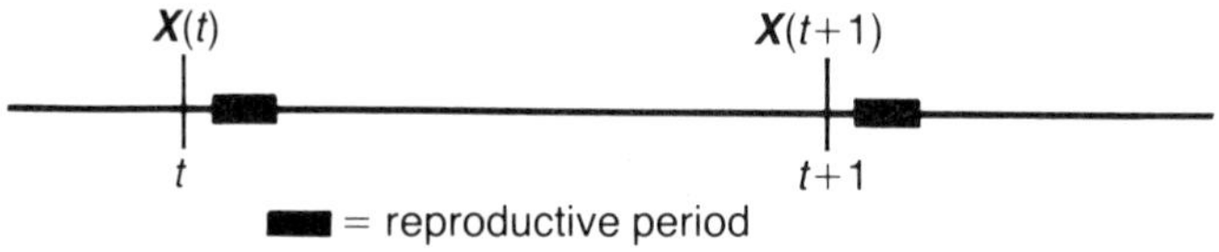

Then $x_0(t + 1) = S_0(M_0x_0(t) + M_1x_1(t) + \ldots + M_nx_n(t))$. Thus we have the formulas

$$\boxed{F_0 = S_0M_0, \quad F_1 = S_0M_1, \ldots, F_n = S_0M_n}$$

Case II We census the population just after reproduction.

$$\begin{aligned} \text{Then } x_0(t + 1) &= M_1x_1(t + 1) + M_2x_2(t + 1) + \ldots + M_nx_n(t + 1) \\ &= M_1S_0x_0(t) + M_2S_1x_1(t) + \ldots + M_nS_{n-1}x_{n-1}(t) \end{aligned}$$

It follows that

$$\boxed{F_0 = S_0M_1, \quad F_1 = M_2S_1, \ldots, F_{n-1} = M_nS_{n-1}, \quad F_n = 0}$$

Note that in both cases we are assuming that the reproductive period is so short that we can ignore mortality during this period. Since the two sets of formulae are so different, it is important that one states precisely when the population is being counted. Unfortunately, this is seldom the case in the literature.

Example 4.3 A population of trout is divided into the four age classes 'egg', 'one year old', 'two year old', and 'three year old'. Two and three year olds produce on the average 100 and 150 eggs each year, but only 3 out of every 100 eggs survive the year. Thus $M_0 = 0, M_1 = 0, M_2 = 100, M_3 = 150$, and $S_0 = 0.03$. If we count the population yearly before reproduction, $F_0 = 0, F_1 = 0$, $F_2 = 3$, and $F_4 = 4.5$. If we census the population after reproduction, we will need to know S_1 and S_2 before we can compute the fecundity constants.

4.4 EIGENVALUE–EIGENVECTOR THEOREMS FOR LESLIE MATRICES

The fact that there are so many zeros in a Leslie matrix might lead one to believe that finding eigenvalues and eigenvectors is easier than in general. Our next two theorems will show that this is indeed the case. In addition, we will give simple conditions on the fecundity rates F_k which will insure the existence of a dominant eigenvalue.

Theorem 4.1 The characteristic polynomial of a Leslie matrix **P** of order $n + 1$ is given by

$$p_n(\lambda) = \lambda^{n+1} - F_0\lambda^n - (S_0F_1)\lambda^{n-1} - (S_0S_1F_2)\lambda^{n-2} - \ldots - (S_0S_1 \cdots S_{n-1}F_n)$$

Proof of Theorem 4.1 We will use induction and expand the determinant of $\lambda\mathbf{I} - \mathbf{P}$ along the bottom row. For $n = 1$,

$$\begin{vmatrix} \lambda - F_0 & -F_1 \\ -S_0 & \lambda \end{vmatrix} = \lambda^2 - F_0\lambda - S_0F_1$$

Assume that the formula for $p_n(\lambda)$ is true and note that a Leslie matrix of order $n + 2$ is of the form

$$\left[\begin{array}{ccc|c} & & & F_{n+1} \\ & & & 0 \\ & \mathbf{P}' & & \vdots \\ \hline 0 & \ldots & S_n & 0 \end{array}\right]$$

where $\mathbf{P}'$ is a Leslie matrix of order $n + 1$. Computing $\lambda\mathbf{I} - \mathbf{P}$ and expanding along the last row,

$$p_{n+1}(\lambda) = S_n \det \mathbf{Q} + \lambda p_n(\lambda)$$

where

$$
\mathbf{Q} = \begin{bmatrix}
\lambda - F_0 & -F_1 & \dots & -F_{n-2} & -F_{n-1} & -F_{n+1} \\
-S_0 & \lambda & \dots & 0 & 0 & 0 \\
\vdots & \vdots & \dots & \vdots & \vdots & \vdots \\
0 & 0 & & -S_{n-2} & \lambda & 0 \\
0 & 0 & & 0 & -S_{n-1} & 0
\end{bmatrix}
$$

The determinant of $\mathbf{Q}$ may be found by expanding down the last column to obtain $\det \mathbf{Q} = (-1)^n\ [(-F_{n+1})(-1)^n\ S_0 S_1 \cdots S_{n-1}]$. Thus we have shown

$$p_{n+1}(\lambda) = \lambda p_n(\lambda) - (S_n S_{n-1} \cdots S_1 S_0) F_{n+1}$$

and the result for $n + 1$ follows upon algebraic simplification.

Corollary 4.1.1 A Leslie matrix has at least one positive real eigenvalue.

Proof of Corollary 4.1.1 If $F_n > 0$, then $p_n(0) = -(S_0 \cdots S_{n-1})F_n < 0$. Since $\lim_{\lambda \to \infty} p_n(\lambda) = +\infty$, we may apply the intermediate value theorem to obtain a zero for $p_n(\lambda)$. If $F_n = 0$, $p_n(\lambda) = \lambda p_{n-1}(\lambda)$ and we may apply induction.

Theorem 4.2 If λ is a non-zero eigenvalue for a Leslie matrix $\mathbf{P}$, then a non-zero eigenvector is given by

$$\mathbf{E}_\lambda = [1, S_0/\lambda, S_0 S_1/\lambda^2, \dots, (S_0 S_1 \cdots S_{n-1})/\lambda^n]$$

Proof of Theorem 4.2 The equation $\mathbf{PX} = \lambda\mathbf{X}$ implies that $S_0 x_0 = \lambda x_1$, $S_1 x_1 = \lambda x_2, \dots, S_{n-1} x_{n-1} = \lambda x_n$. If we set $x_0 = 0$, then all x_is are zero and the eigenvector has reduced to the zero vector. If we set $x_0 = 1$, then $x_1 = S_0/\lambda$, $x_2 = S_1 x_1/\lambda = S_0 S_1/\lambda^2$, and so on.

The eigenvalue components in Theorem 4.2 can be generated from the recursion $x_{k+1} = S_k x_k/\lambda$ with $x_0 = 1$. When combined with a subroutine that finds all roots of a real polynomial, Theorems 4.1 and 4.2 provide the formulas needed to give complete eigenanalysis of a Leslie matrix. You will be asked to write the necessary program in the exercises. The use of such a program is illustrated in our next example.

Example 4.4 In a 1971 paper [4], A. L. Jensen applied the Leslie matrix model to a population of brook trout in Hunt Creek, Michigan. The population was divided into five year-classes (fingerlings, yearlings, etc.) and the Leslie matrix, with entries rounded off to two significant digits, is given by

$$\mathbf{P} = \begin{bmatrix} 0 & 0 & 37 & 64 & 82 \\ 0.06 & 0 & 0 & 0 & 0 \\ 0 & 0.34 & 0 & 0 & 0 \\ 0 & 0 & 0.16 & 0 & 0 \\ 0 & 0 & 0 & 0.08 & 0 \end{bmatrix}$$

The characteristic polynomial $\lambda^5 - 0.7548\,\lambda^2 - 0.208896\,\lambda - 0.0214118$ has roots $\lambda = 0.995403, -0.138259 \pm 0.095100\,i$, and $-0.359442 \pm 0.79667\,i$. Thus the dominant eigenvalue is $\lambda_0 = 0.995403$ and $|\lambda_1| = 0.874$. Using Theorem 4.2, we find that $\mathbf{S} = [0.922, 0.056, 0.019, 0.003, 0.0002]$ is the stable age distribution. Since there are errors in estimating the survival and fecundity rates, we may conclude that $\lambda_0 \approx 1$. When $\lambda_0 = 1$, we say the population is *stationary*. The stable age distribution will be reached slowly, however, since $|\lambda_1/\lambda_0| = 0.878$.

If the survival rate for the fingerlings could be increased by 20%, $S_0 = 1.2(0.06) = 0.072$, and the new dominant eigenvalue is $\lambda_0 = 1.05284$. Eventually the population would increase in each age class by 5.28% each year.

Finally, suppose that all fish in the last two age classes are captured before reproduction. We now have $F_3 = F_4 = 0$, $S_3 = 0$, and new characteristic polynomial $\lambda^5 - 0.7548\,\lambda^2 = \lambda^2\,(\lambda^3 - 0.7548)$. The dominant eigenvalue is now $\lambda_0 = 0.91049$ and the population will slowly die out.

In most populations, there will be at least *two consecutive age classes* that are fertile. The next theorem establishes that under these common circumstances a *positive real dominant eigenvalue* always exists.

Theorem 4.3 Suppose that there is a k with both F_k and F_{k+1} non-zero. Then the Leslie matrix **P** has a real positive dominant eigenvalue.

Proof of Theorem 4.3 First note that $p_n(\lambda) = 0$ implies

$$\lambda^{n+1} = F_0\lambda^n + S_0F_1\lambda^{n-1} + S_0S_1F_2\lambda^{n-2} + \ldots + (S_0S_1 \cdots S_{n-1})F_n$$

Hence, a non-zero eigenvalue λ will satisfy the identity

$$f(\lambda) = F_0/\lambda + (S_0F_1)/\lambda^2 + (S_0S_1F_2)/\lambda^2 + \ldots + (S_0 \cdots S_{n-1})F_n/\lambda^{n+1}$$
$$= 1$$

Note that $f'(\lambda) < 0$, $\lim_{\lambda \to +\infty} f(\lambda) = 0$, and $\lim_{\lambda \to 0^+} f(\lambda) = +\infty$. The graph of $f(\lambda)$ is shown in Fig. 4.2.

Hence there exists a unique $\lambda_0 > 0$ with $f(\lambda_0) = 1$. Since $f'(\lambda_0) < 0$, λ_0 is of multiplicity one. Let λ_j be any other non-zero eigenvalue and write $1/\lambda_j = r\,e^{i\theta}$. Although no other positive eigenvalues exist, it is still conceivable that λ_j is negative or complex with $|\lambda_j| > \lambda_0$. From De Moivre's theorem,

$$1/\lambda_j^m = r^m\,e^{im\theta} = r^m \cos m\theta + i\,r^m \sin m\theta$$

and so

$$1 = f(\lambda_j) = F_0(re^{i\theta}) + S_0F_1(r^2 e^{i2\theta}) + \ldots + S_0 \cdots S_{n-1}F_n(r^{n+1}\ e^{i(n+1)\theta})\ .$$

Taking real parts, we obtain

$$1 = F_0 r \cos\theta + S_0F_1r^2 \cos 2\theta + \ldots + S_0S_1 \cdots S_{n-1}F_n r^{n+1} \cos(n+1)\theta$$

By hypothesis we have F_k and F_{k+1} non-zero for some k. Also note that $\cos k\theta$ and $\cos(k+1)\theta$ are not both 1. Otherwise $k\theta, (k+1)\theta$, and therefore θ would be multiples of 2π. This would imply that $\lambda_j = 1/r$, real and positive. Hence we may conclude that there is an m with

$$S_0S_1 \cdots S_{m-2}F_{m-1} \cos m\theta < S_0S_1 \cdots S_{m-2}F_{m-1}$$

If $|\lambda_j| = 1/r \geqslant \lambda_0$, then $r \leqslant 1/\lambda_0$ and

$$\begin{aligned} 1 = f(\lambda_0) &\geqslant rF_0 + r^2S_0F_1 + \ldots + S_0S_1 \cdots S_{m-2}F_{m-1}\ r^m + \ldots \\ &> F_0 r \cos\theta + \ldots + S_0S_1 \cdots S_{m-2}F_{m-1}\ r^m \cos m\theta + \ldots \\ &= f(\lambda_j) = 1. \end{aligned}$$

This contradiction implies that $|\lambda_j| < \lambda_0$, and the proof is complete.

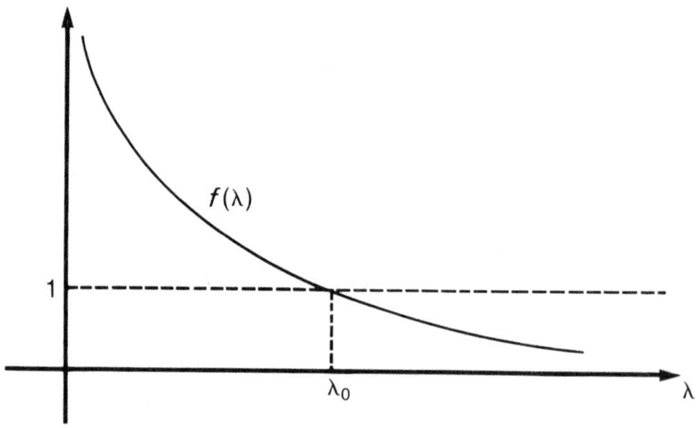

Fig. 4.2: The graph of $f(\lambda)$

The following is a generalization of Theorem 4.3 whose proof is a modification of the one just given and is outlined in the exercises. Let $k_1, k_2, \ldots, k_m$ be the column numbers where the fecundities are non-zero. Then if the greatest common divisor of $k_1, \ldots, k_m$ is 1, **P** will have a real positive dominant eigenvalue. As an illustration, let **P** be the Leslie matrix

$$\begin{bmatrix} 0 & 0 & 14 & 0 & 20 \\ 0.8 & 0 & 0 & 0 & 0 \\ 0 & 0.75 & 0 & 0 & 0 \\ 0 & 0 & 5/6 & 0 & 0 \\ 0 & 0 & 0 & 0.8 & 0 \end{bmatrix}$$

Theorem 4.3 does not apply, but $\{k\colon F_{k+1} \neq 0\} = \{3, 5\}$ and the greatest common divisor of 3 and 5 is 1. The theorem then states that a real positive dominant eigenvalue exists. (The actual eigenvalues of the matrix **P** are $\lambda = 2.16237, 0.0550 \pm 0.9640\,i$, and $-1.13121 \pm 1.60435\,i$.)

Theorem 4.3 and its generalization can be deduced from a general theorem in matrix algebra (see [2] for a proof).

Perron–Frobenius Theorem Let **A** be an $n \times n$ matrix with $a_{ij} \geqslant 0$ for all i and j. Then if $\mathbf{A}^m$ has all positive entries for some m, **A** has a real positive dominant eigenvalue of multiplicity one.

The exercises sections contain several applications of the Leslie matrix model taken from the literature.

EXERCISES

Programming exercises

The programs requested in Exercises 1 and 2 must be used in solving several of the exercises that follow.

1 Write a program that inputs survival rates, fecundity rates, and the initial population **X**(0) and outputs population projections.

2. Make use of a software subroutine that finds all zeros of a real polynomial (see p. 38) and use Theorems 4.1 and 4.2 to find all eigenvalues and the stable age distribution for a Leslie matrix **P**.

Part A

3 The Leslie matrix for an insect population is

$$\mathbf{P} = \begin{bmatrix} 0 & 20 & 0 \\ 0.1 & 0 & 0 \\ 0 & 0.2 & 0 \end{bmatrix}$$

The age classes are 0–1 week, 2–3 weeks, and 4–5 weeks, and only insects 2–3 weeks old can reproduce.

(a) If **X**(0) = [0, 100, 0] specifies the initial age structure, project the population forward 2 months.

(b) Show that $\mathbf{A}^3 = 2\mathbf{A}$ and $\mathbf{A}^4 = 2\mathbf{A}^2$. Conclude that the population in each age class doubles every month.

4 The growth of many fish populations, such as salmon or striped bass, is characterized by very high fecundities and extremely small survival rates among the newborns. Suppose that such a population is divided 0-year-olds (eggs), 1-year-olds, 2-year-olds, and 3-year-olds, and the Leslie matrix is given by

$$\mathbf{P} = \begin{bmatrix} 0 & 0 & 0 & 1000 \\ 0.005 & 0 & 0 & 0 \\ 0 & 0.6 & 0 & 0 \\ 0 & 0 & 0.7 & 0 \end{bmatrix}$$

(a) Project the population forward 5 years given that $\mathbf{X}(0) = [1000, 0, 0, 0]$.
(b) Show that $\mathbf{A}^4 = 2.1\ \mathbf{I}$. Conclude that the population increases in each age class by 210% every four years.

5 If $\mathbf{P}$ is the Leslie matrix

$$\begin{bmatrix} 0 & F_1 & 0 \\ S_0 & 0 & 0 \\ 0 & S_1 & 0 \end{bmatrix}$$

show that $\mathbf{P}^3 = (S_0 F_1)\ \mathbf{P}$ and $\mathbf{P}^4 = (S_0 F_1)\ \mathbf{P}^2$. If $\mathbf{X}_n$ denotes the population structure after n time periods, find $\mathbf{X}_n$ in terms of $\mathbf{X}_1$ and $\mathbf{X}_2$.

6 If $\mathbf{P}$ is the Leslie matrix

$$\begin{bmatrix} 0 & 0 & F_2 \\ S_0 & 0 & 0 \\ 0 & S_1 & 0 \end{bmatrix}$$

(a) Show that $\mathbf{P}^3 = (S_0 S_1 F_2)\ \mathbf{I}$
(b) If $S_0 S_1 F_2 < 1$, what can you predict about future generations?
(c) If $F_2 = 4$ and $S_1 = 0.7$, find S_0 so that the eventual population growth is 50% every three time periods.

7 For a 2 X 2 Leslie matrix, determine conditions on F_0, F_1, and S_0 that will insure a dominant eigenvalue $\lambda_0 > 1$.

For each of the following Leslie matrices, (a) determine λ_0 and the stable age structure $\mathbf{S}$, and (b) find $|\lambda_1 / \lambda_0|$.

8 $\begin{bmatrix} 0 & 1 & 2 \\ 0.8 & 0 & 0 \\ 0 & 0.7 & 0 \end{bmatrix}$ **9** $\begin{bmatrix} 0 & 2 & 4 & 6 \\ 0.2 & 0 & 0 & 0 \\ 0 & 0.4 & 0 & 0 \\ 0 & 0 & 0.6 & 0 \end{bmatrix}$

10 The following information is provided by N. Leader-Williams [6] on the population dynamics of reindeer herds on South Georgia Island in the Southern Atlantic Ocean. Ten Norwegian reindeer were introduced to the island by whalers in 1911. For over forty years, the herd thrived with population reaching about 3000 in 1958. Since then the population has been declining. Single calves are born in November and survive the next year with probability 0.71. Conception can take place at age 1½ and the pregnancy rate is 90%. The survival rate for yearlings and adults is 0.68 and adult females can live to age 12 years.

(a) If we count the population after reproduction (i.e. in November), construct an appropriate Leslie matrix **P**. Assume male–female births are equiprobable.
(b) Find the dominant eigenvalue λ_0 and the stable age distribution **S**.

11 The survival rate S_0 for newborns in an animal population is dependent on the severity of winter. If **P** is the Leslie matrix

$$\begin{bmatrix} 0 & 2S_0 & 4S_0 \\ S_0 & 0 & 0 \\ 0 & 0.8 & 0 \end{bmatrix}$$

find the smallest S_0 so that the population can maintain itself over time.

12 An insect population breeds every two weeks but death is certain after six weeks. The species is mature at age 2 weeks and produces 100 eggs. Insects aged 4 weeks produce 150 eggs. In addition, the survival rates are $S_0 = 0.05$ and $S_1 = 0.2$.

(a) If the population is counted after reproduction, find an appropriate Leslie matrix **P** and the dominant eigenvalue λ_0.
(b) In an attempt to control the population, an insecticide is used which will cut survival rates in half. Will the spraying program be successful?

13 (Exercise is based on [9], p. 59, exercise 9.) An animal population grows according to matrix $\mathbf{P}_1$ during six winter months and according to matrix $\mathbf{P}_2$ during six summer months.

$$\mathbf{P}_1 = \begin{bmatrix} 0 & 2 & 3 \\ 1/2 & 0 & 0 \\ 0 & 2/3 & 0 \end{bmatrix} \qquad \mathbf{P}_2 = \begin{bmatrix} 0 & 3 & 4 \\ 1/3 & 0 & 0 \\ 0 & 3/4 & 0 \end{bmatrix}$$

(a) Assuming that the stable age distribution has been reached, how fast is the population increasing?

The number of animals is reaching plague proportions and the government is about to initiate an extermination program. Unfortunately, the method involved is expensive, but will kill 50% of the animals under 6 months, $33\frac{1}{3}$% of the animals aged 6–12 months, and 25% of the animals over 12 months within a few days after the annual application.

(b) Should the scheme be applied in autumn or spring?

14 In 1945, as an illustration of his new model, P. H. Leslie used the 1939 data of King on a laboratory population of the brown rat to construct the entries in the table below. Age is measured in months.

Age group	0–	1–	2–	3–	4–	5–	–6
S_k	0.94697	0.99665	0.99926	0.99899	0.99863	0.99817	0.99753
F_k	0	0	0.3964	1.4939	2.1777	2.5250	2.6282
Age group	7–	8–	9–	10–	11–	12–	13–
S_k	0.99667	0.99553	0.99399	0.99196	0.98926	0.98572	0.98107
F_k	2.6749	2.6018	2.4419	2.1865	1.9044	1.7259	1.4918
Age group	14–	15–	16–	17–	18–	19–	20–
S_k	0.97511	0.96748	0.95797	0.94631	0.93247	0.91649	0
F_k	1.2415	0.9522	0.7141	0.4618	0.2518	0.0901	0.0035

(a) Find the dominant eigenvalue λ_0 and the stable age distribution.
(b) Find the next largest eigenvalue λ_1.
(c) Starting with 1000 newborns only, project the population forward 20 generations.

15 In the late 1970s, the Leslie matrix model was used to make predictions on the fate of the Hudson River striped bass population. Adults live in the Atlantic Ocean, but return yearly to swim up the Hudson to spawn. Since industrial development along the Hudson poses significant pollution problems (e.g. it increases water temperature which may effect fecundity and survival rates), it is important to know what are the effects on this popular sports fish.

The population is divided into the 16 age classes 0-year-old (eggs), 1-year-old (larvae), 2-year-old, . . . , 15-year-old. Ages 5–15 are mature and ages 3–15 are subject to fishing. Survival rates shown in the table take into account both natural and fishing mortality. Best estimates are shown below.

Age class	0	1	2	3	4	5
S_k	2.121×10^{-5}	0.3965	0.6000	0.8000	0.6387	0.5688
F_k	0	0	0	0	0	80 110

Age class	6	7	8	9	10
S_k	0.5688	0.5688	0.5688	0.5688	0.5688
F_k	162 700	212 700	267 900	326 400	386 000

Age class	11	12	13	14	15
S_k	0.5688	0.5688	0.5688	0.5688	0.5688
F_k	444 500	499 700	549 600	592 200	592 200

The best estimate of the present population (in thousands) is

$$\mathbf{X}(0) = [5.21 \times 10^7, 1100, 443, 266, 213, 136, 77, 43.7, 24.8, \\ 14.1, 7.99, 4.53, 2.57, 1.46, 0.829, 1.087]^T.$$

(a) Find the dominant eigenvalue λ_0 and the stable age distribution under the present conditions.
(b) Assuming ecological conditions are unchanged, project the population forward until the stable age distribution is 'reached' to two correct significant digits.
(c) Suppose that further industrial development decreases the survival rate for eggs by 25%, of larvae by 15%, and the fecundity of mature adults by 10%. Perform the eigenvalue analysis and forecast the fate of the population. How many years will it take to cut the *fishable population* in half?

16 In human demography the age categories are usually 0–4 years, 5–9 years, . . . , 80–84 years, and 85+ years. By age, we mean 'age last birthday'. The unit for the projection is therefore 5 years. Shown below are survival and fecundity rates for US females as of 1967.

Age class	0–4	5–9	10–14	15–19	20–24	25–29	30–34
S_k	0.99694	0.99842	0.99783	0.99671	0.99614	0.99496	0.99247
F_k	0	0.00105	0.08203	0.28849	0.37780	0.26478	0.14055

Age class	35–39	40–44	45–49	50–54	55–59	60–64	65–69
S_k	0.98875	0.98305	0.97473	0.96297	0.94624	0.91713	0.87113
F_k	0.05857	0.01344	0.00081	0	0	0	0

Age class	70–74	75–79	80–84
S_k	0.80572	0.70302	0.80246
F_k	0	0	0

(a) Find the dominant eigenvalue λ_0 and the stable age distribution **S**.

The age distribution in June 1967, was

$$[0.093, 0.101, 0.097, 0.087, 0.074, 0.060, 0.055, 0.058, 0.063, 0.060, 0.055, 0.049, 0.042, 0.035, 0.029, 0.022, 0.013, 0.007]$$

and the total female population was 101 169 000 (Fig. 4.3).

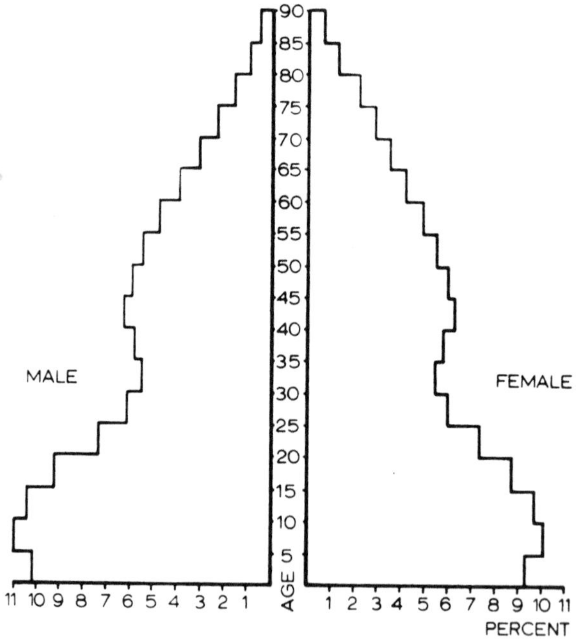

Fig. 4.3: Population pyramid for the United States, 1967

* From *Population: Facts and Methods of Demography*, by N. Keyfitz and W. Flieger, W. H. Freemand and Co., Copyright (c) 1971. All rights reserved.

(b) Make population projections for June 1982, June 1987, June 1992, and June 1997.

(c) How quickly will the stable age distribution be 'reached' (to two correct significant digits)?

(d) How does the college age population (ages 15–24) change?
(e) How does the retirement age population (65+) change?

In 1967, the ratio of males to females in the various age categories was specified by the vector (Fig. 4.4)

$$\mathbf{R} = [1.04224, 1.03642, 1.03240, 1.01423, 0.93519, 0.96581, 0.95117, 0.95116, 0.94511, 0.93816, 0.93898, 0.92344, 0.89365, 0.83488, 0.76028, 0.72202, 0.67962, 0.61348]$$

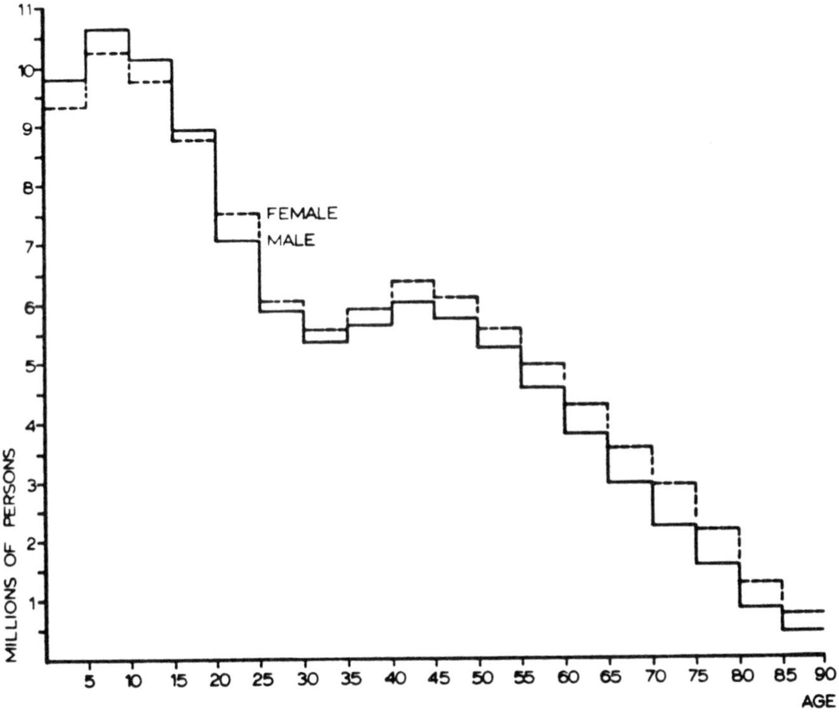

Fig. 4.4: Male and female populations by age, United States, 1967

* From *Population: Facts and Methods of Demography*, by N. Keyfit and W. Flieger, W. H. Freeman and Co., Copyright (c) 1971. All rights reserved.

(f) Answer questions (b)–(e) for males.

Part B

17 If $\mathbf{X}(t+1) = \mathbf{P}\,\mathbf{X}(t)$, find a formula for $\mathbf{X}(0)$ in terms of $\mathbf{X}(n)$.

18 If $\mathbf{P}$ is a Leslie matrix, find conditions on the survival and fecundity rates that will insure that $\mathbf{P}$ has an inverse.

19 Prove that $\mathbf{P}$ has a real domonant eigenvalue $\lambda_0 > 1$ if and only if

$$F_0 + S_0F_1 + S_0S_1F_2 + \ldots + S_0S_1 \cdots S_{n-1}F_n > 1.$$

20 Let $l_0 = 1, l_1 = S_0, l_2 = S_0 S_1, \ldots, l_n = S_0 S_1 \cdots S_{n-1}$ and $l_{n+1} = 0$.

(a) Show that l_k is the probability of surviving to age class k.

(b) Let $p_k = l_k - l_{k+1}$. Show that p_k is the probability of surviving to age class k *only*.

(c) Simplify $\sum_{k=0}^{n} p_k(F_0 + \ldots + F_k)$. What is the interpretation of this sum?

21 Let **P** be a Leslie matrix of order $n + 1$ and let $F_k = F_{k+1} = \ldots = F_n = 0$. Let $\mathbf{P}'$ be the Leslie sub-matrix of order $k + 1$. Show that the non-zero eigenvalues of **P** are the non-zero eigenvalues of $\mathbf{P}'$.

22 Let **P** be a Leslie matrix and $k_1, k_2, \ldots, k_m$ the column numbers with non-zero fecundities. Assume the greatest common divisor of $k_1, \ldots, k_m$ is 1.

(a) Show that (as in the proof of Theorem 4.3) it suffices to find a k with $F_k > 0$ and $\cos k\theta < 1$ in order to establish that **P** has a real positive dominant eigenvalue.

(b) By a standard theorem in number theory, there are integers $z_1, \ldots, z_m$ with $k_1 z_1 + k_2 z_2 + \ldots + k_m z_m = 1$. If $\cos k_i\theta = 1$ for $i = 1, \ldots, m$, conclude that θ is a multiple of 2π.

(c) Conclude from (b) that λ_j is positive and real, a contradiction.

REFERENCES

[1] H. Bernadelli, 'Population Waves', *J. Burma Res. Soc.* (1941), **31**, 1–18.

[2] A. Brauer, 'A New Proof of the Theorems of Perron and Frobenius on Non-negative Matrices', *Duke Math. J.* (1957), **24**, 367–378.

[3] C. Fowler and T. Smith, 'Characterizing Stable Populations: an Application to the African Elephant Population', *J. Wildlife Management* (1973), **37**(4), 513–523.

[4] A. L. Jensen, 'The Effect of Increased Mortality on the Young in a Population of Brook Trout, a Theoretical Analysis', *Trans. Amer. Fish. Soc.* (1971), **100**(1), 456–459.

[5] N. Keyfitz, *An Introduction to the Mathematics of Populations*, Addison-Wesley, 1968.

[6] N. Leader-Williams, 'Population Dynamics and Mortality of Reindeer Introduced into South Georgia', *J. Wildlife Management* (1980), **44**(3), 640–665.

[7] P. H. Leslie, 'On the Use of Matrices in Certain Population Mathematics', *Biometrika* (1945), **33**, 183–212.

[8] E. G. Lewis, 'On the Generation and Growth of a Population', *Sankya* (1943), **6**, 93–96.

[9] J. H. Pollard, *Mathematical Models for the Growth of Human Populations*, Cambridge University Press, 1973.

CHAPTER 5

Generalizations of the Leslie Matrix Model

Since its introduction in the 1940s, the Leslie matrix model as been generalized and modified in a large number of ways. In this chapter, we will study three such modifications. In the first model, members of the last age class, rather than being removed from the population, can survive subsequent time periods and continue to reproduce. Next we will show how the basic model can be adapted to keep track of both sexes. Finally, we will present a forest harvesting model in which trees are classified by size rather than age.

5.1 SURVIVAL IN THE LAST AGE CLASS

Suppose that the members of the last age class do not die, as the basic Leslie matrix model supposes, but can live and perhaps reproduce for some years. As an example, consider the blue whale population. The reproductive cycle of the blue whale (see page 50) suggests that we divide the population into age classes of length two years. Since a member of the species may live fifty years or more and since a female will continue to reproduce, this would require a regular Leslie matrix of size at least 25×25. On the other hand, individuals 12 years old and older are 'essentially the same'. They survive a given two-year period with the same probability (about 0.8) and continue to reproduce with the same regularity ($F \approx 0.45$). This general situation is true for a number of large mammals. For such populations, the last equation $x_n(t+1) = S_{n-1}\, x_{n-1}(t)$ is replaced by

$$x_n(t+1) = S_{n-1}\, x_{n-1}(t) + S_n x_n(t)$$

and is illustrated in Fig. 5.1.

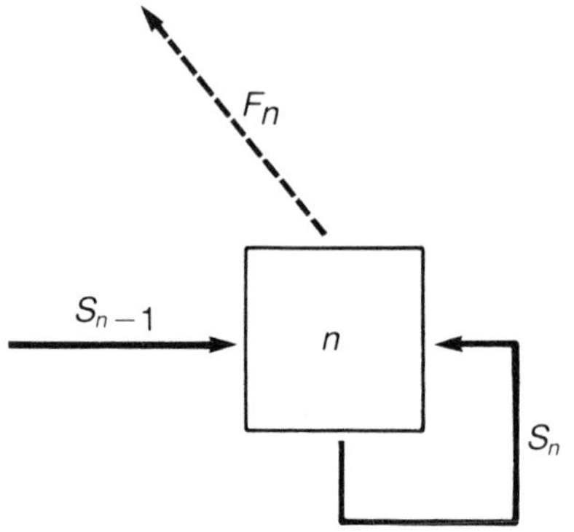

Fig. 5.1: Survival in the last age class

The new eigenvalue–eigenvector formulas are closely related to those in Theorems 4.1 and 4.2.

Theorem 5.1 For the Leslie matrix model with feedback probability S_n, the characteristic polynomial $q_n(\lambda)$ is given by

$$q_n(\lambda) = p_n(\lambda) - S_n\, p_{n-1}(\lambda)$$

where $p_n(\lambda)$ is as in Theorem 4.1.

Proof of Theorem 5.1 We can proceed as in the proof of Theorem 4.1 and expand $\det(\lambda\mathbf{I} - \mathbf{P})$ along the last row to obtain

$$q_n(\lambda) = \det(\lambda\mathbf{I} - \mathbf{P}) = (\lambda - S_n)p_{n-1}(\lambda) + S_0 \cdots S_{n-1} F_n$$

But $\lambda p_{n-1}(\lambda) + S_0 \ldots S_{n-1} F_n = p_n(\lambda)$. Hence $q_n(\lambda) = p_n(\lambda) - S_n p_{n-1}(\lambda)$.

Theorem 5.2 If λ is an eigenvalue of $\mathbf{P}$ and $\lambda \neq 0, S_n$, then a corresponding eigenvector is given by

$$\mathbf{E}_\lambda = [1, S_0/\lambda, S_0 S_1/\lambda^2, \ldots, S_0 \cdots S_{n-2}/\lambda^{n-1}, S_0 \cdots S_{n-1}/[\lambda^{n-1}(\lambda - S_n)]]$$

Note that the eigenvector is generated by the recursion $x_0 = 1$, $x_{k+1} = S_k x_k/\lambda$ for $k = 0, \ldots, n-2$, and $x_n = S_{n-1} x_{n-1}/(\lambda - S_n)$. One of the exercises will ask that you modify the computer programs that were written in Chapter 4 to perform the eigenvalue analysis. Note that when $S_n = 0$, Theorems 5.1 and 5.2 reduce to Theorems 4.1 and 4.2. These new formulas are illustrated in the following two examples.

Example 5.1 The Leslie matrix for the American bison is given by

$$\mathbf{P} = \begin{bmatrix} 0 & 0 & 0.42 \\ 0.60 & 0 & 0 \\ 0 & 0.75 & 0.95 \end{bmatrix}$$

The female population is divided into calves, yearlings, and adults (age 2 years or more). Adults survive an additional year with probability 0.95 and reproduce with the same regularity. The characteristic polynomial is $\lambda^3 - 0.95\lambda^2 - 0.189$ with roots $\lambda = 1.1048$ and $-0.0774 \pm 0.40629\, i$. Thus $|\lambda_1| = 0.4136$ and $|\lambda_1/\lambda_0| = 0.374$. This indicates that the stable age distribution

$$\mathbf{S} = [0.2396, 0.1301, 0.6303]$$

should be reached quickly. To demonstrate this, suppose that a herd is started with 100 adult females (and an appropriate number of males). We obtain the population projections in Table 5.1. After only five years the stable age distribution is reached to two correct decimal places.

Table 5.1

Year	Calves	Yearlings	Adults
0	0	0	100
1	42	0	95
2	40	25	90
3	38	24	105
4	44	23	117
5	49	26	129

Example 5.2 Based on vital statistics from the 1930s, M. B. Usher ([6], p. 176) constructed the following Leslie matrix for the female blue whale population.

$$\mathbf{P} = \begin{bmatrix} 0 & 0 & 0.19 & 0.44 & 0.50 & 0.50 & 0.45 \\ 0.87 & 0 & 0 & 0 & 0 & 0 & 0 \\ 0 & 0.87 & 0 & 0 & 0 & 0 & 0 \\ 0 & 0 & 0.87 & 0 & 0 & 0 & 0 \\ 0 & 0 & 0 & 0.87 & 0 & 0 & 0 \\ 0 & 0 & 0 & 0 & 0.87 & 0 & 0 \\ 0 & 0 & 0 & 0 & 0 & 0.87 & 0.80 \end{bmatrix}$$

$$\mathbf{Y}_\lambda = y_0\ [1, \bar{S}_0/\lambda, \ldots, \bar{S}_0 \cdots \bar{S}_{n-2}/\lambda^{n-1}, \bar{S}_0 \cdots \bar{S}_{n-1}/[\lambda^{n-1}(\lambda - \bar{S}_n)]]$$

with $y_0 = (\bar{F}_0 x_0 + \bar{F}_1 x_1 + \ldots + \bar{F}_n x_n)/\lambda$.

These new eigenvalue–eigenvector formulas are used in our next example.

Example 5.3 The matrix **A** of the two-sex population model for the American bison is shown below.

$$\mathbf{A} = \left[\begin{array}{ccc|ccc} 0 & 0 & 0.42 & 0 & 0 & 0 \\ 0.6 & 0 & 0 & 0 & 0 & 0 \\ 0 & 0.75 & 0.95 & 0 & 0 & 0 \\ \hline 0 & 0 & 0.48 & 0 & 0 & 0 \\ 0 & 0 & 0 & 0.6 & 0 & 0 \\ 0 & 0 & 0 & 0 & 0.75 & 0.95 \end{array}\right]$$

Note then that the survival rates for males and females are identical, but male births are more frequent. By Theorem 5.3, **A** has characteristic polynomial $\lambda^2(\lambda - 0.95)(\lambda^3 - 0.95\lambda^2 - 0.189)$. The dominant eigenvalue is again $\lambda_0 = 1.10483$ and the stable age distribution is

$$\mathbf{S} = [0.112, 0.061, 0.294, 0.128, 0.069, 0.336].$$

But note that λ_1 is now 0.95 and so $|\lambda_1/\lambda_0| = 0.860$. We therefore expect **S** to be reached much more slowly than in Example 5.1. If, for example, $\mathbf{X}(0) = [0, 0, 100, 0, 0, 20]$, then it takes more than 25 years for the stable age distribution to be reached to two correct decimal places.

5.3 A FOREST HARVESTING MODEL

A population of trees is divided into $n + 1$ size classes $C_0, C_1, \ldots, C_n$, and we let $x_i(t)$ = number of trees in size category C_i at time t. We wish to determine the forest structure and harvesting policy that will result in large sustained yields and which, at the same time, will provide long range conservation of the resource. In our model, we will make the following assumptions:

1. *A tree may advance at most one size class during the unit period from t to $t + 1$.*

Let b_k = probability that a tree enlarges from size class C_k to size class C_{k+1} and let $a_k = 1 - b_k$. We will either assume that no deaths occur or alternately that a dead tree will always be a part of the harvest. Hence $x_k(t + 1) = a_k\, x_k(t) + b_{k-1} x_{k-1}(t)$. The recruitment scheme is then specified by the matrix

$$\mathbf{A} = \begin{bmatrix} a_0 & 0 & 0 & \dots & 0 & 0 \\ b_0 & a_1 & 0 & \dots & 0 & 0 \\ 0 & b_1 & a_2 & \dots & 0 & 0 \\ \vdots & \vdots & \vdots & & \vdots & \vdots \\ 0 & 0 & 0 & \dots & b_{n-1} & 1 \end{bmatrix}$$

In order that all size classes be attainable, we assume that $0 < b_k \leqslant 1$.

2. *Regeneration can occur only when trees are thinned from the forest.*

Let c_k = number of trees in size class C_0 which develop over the next time period when a tree from size class C_k is removed. We will assume that $c_0 < 1$ and $c_n > 1$. Regeneration is then specified by the matrix

$$\mathbf{C} = \begin{bmatrix} c_0 & c_1 & c_2 & \dots & c_n \\ \hline & & \mathbf{0} & & \end{bmatrix}$$

3. *The enumeration* $\mathbf{X}(t)$ *takes place right before harvesting.*

Let h_k be the percentage of size class C_k left after harvesting and let

$$\mathbf{H} = \begin{bmatrix} h_0 & 0 & 0 & \dots & 0 \\ 0 & h_1 & 0 & \dots & 0 \\ 0 & 0 & h_2 & \dots & 0 \\ \vdots & \vdots & \vdots & & \vdots \\ 0 & 0 & 0 & \dots & h_n \end{bmatrix}$$

H is called a *harvesting matrix* and specifies the harvesting policy that will be used again and again. As illustrated in Fig. 5.3, we wish to determine $\mathbf{X}(t+1)$ in terms of $\mathbf{X}(t)$.

It follows that

$$\boxed{\mathbf{X}(t+1) = (\mathbf{AH} + \mathbf{C}[\mathbf{I} - \mathbf{H}])\mathbf{X}(t)}$$

If we let w_k = value of a tree in size class C_k, then the total value of the harvest is given by

$$T = w_0(1 - h_0)x_0 + \dots + w_n(1 - h_n)x_n$$

We might measure w_k in dollars or perhaps cubic feet of lumber. Finally, we make the following conservation assumption.

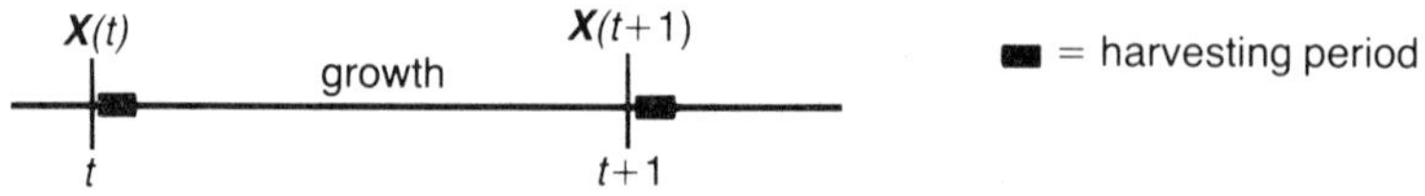

Before thinning	After thinning	After the growth period
x_0	$h_0 x_0$	$a_0(h_0 x_0) + \sum_{k=0}^{n} c_k (1 - h_k) x_k$
x_1	$h_1 x_1$	$b_0(h_0 x_0) + a_1(h_1 x_1)$
x_2	$h_2 x_2$	$b_1(h_1 x_1) + a_2(h_2 x_2)$
$\vdots$	$\vdots$	$\vdots$
x_n	$h_n x_n$	$b_{n-1}(h_{n-1} x_{n-1}) + h_n x_n$

Fig. 5.3: The population is counted before the harvest.

4. *After harvesting and after the growth period, the forest is returned to its prior state, i.e.* $\mathbf{X}(t+1) = \mathbf{X}(t)$.

We wish then to find a vector $\mathbf{X}$ and a harvesting matrix $\mathbf{H}$ so that $\mathbf{X} = (\mathbf{AH} + \mathbf{C}[\mathbf{I} - \mathbf{H}])\,\mathbf{X}$. M. B. Usher [7] has remarked that thinning operations will remove a more or less equal proportion of all age classes in the forest. We will then assume that $\mathbf{H} = h\,\mathbf{I}$ for $0 < h < 1$. (The more general case is presented in Chapter 6.)

We must solve the equation $[h\,\mathbf{A} + (1-h)\,\mathbf{C}]\,\mathbf{X} = \mathbf{X}$ for h and $\mathbf{X}$. Hence $h(\mathbf{A} - \mathbf{C})\mathbf{X} = \mathbf{X} - \mathbf{CX} = (\mathbf{I} - \mathbf{C})\mathbf{X}$. Now $\mathbf{I} - \mathbf{C}$ is invertible. In fact $(\mathbf{I} - \mathbf{C})^{-1} = \mathbf{I} + (1 - c_0)^{-1}\,\mathbf{C}$. Letting $\lambda = 1/h$, it follows that

$$\boxed{(\mathbf{I} + \frac{1}{1 - c_0}\mathbf{C})\,(\mathbf{A} - \mathbf{C})\,\mathbf{X} = \lambda\,\mathbf{X}}$$

Thus $\lambda = 1/h$ *is an eigenvalue and* $\mathbf{X}$ *is an eigenvector for the matrix* $\mathbf{B} = (\mathbf{I} + (1 - c_0)^{-1}\mathbf{C})\,(\mathbf{A} - \mathbf{C})$. Since $0 < h < 1$, $\lambda > 1$. Usher [5] has shown that this problem has a unique solution $\lambda > 1$ with each $x_i \geqslant 0$. The model is applied to a pine forest in our final example.

Example 5.4 The Scots pine forest at Corrour, Inverness-shire, Scotland is divided into the six size classes shown in Table 5.2.

The recruitment probabilities are $b_0 = 0.28$, $b_1 = 0.31$, $b_2 = 0.25$, $b_3 = 0.23$ and $b_4 = 0.37$. In addition, $c_0 = c_1 = c_2 = 0$, $c_3 = 3.6$, $c_4 = 5.1$, and $c_5 = 7.5$. The matrix $\mathbf{B} = (\mathbf{I} + \mathbf{C})\,(\mathbf{A} - \mathbf{C})$ is given by

Table 5.2

	Class 0	Class 1	Class 2	Class 3	Class 4	Class 5
Girth (in.)	15–21	21–27	27–33	34–40	40–46	⩾ 46
Volume (ft³/tree)	2.9	7	13	21	31	43

$$\mathbf{B} = \begin{bmatrix} 0.72 & 0 & 0.90 & 0.345 & 0.888 & 0 \\ 0.28 & 0.69 & 0 & 0 & 0 & 0 \\ 0 & 0.31 & 0.75 & 0 & 0 & 0 \\ 0 & 0 & 0.25 & 0.77 & 0 & 0 \\ 0 & 0 & 0 & 0.23 & 0.63 & 0 \\ 0 & 0 & 0 & 0 & 0.37 & 1 \end{bmatrix}$$

The power method, for example, gives $\lambda = 1.2042662$ as the dominant eigenvalue with corresponding eigenvector

$$\mathbf{S} = [0.4218, 0.2297, 0.1567, 0.0902, 0.0361, 0.0655]$$

giving the stable size distribution. In 1966, the population was $\mathbf{X}(0) = [4461, 2926, 1086, 222, 27, 2]$ and the forest is thinned every six years. Applying the equation $\mathbf{X}(t+1) = \mathbf{P}\,\mathbf{X}(t)$ where

$$\mathbf{P} = h\mathbf{A} + (1-h)\mathbf{C} = \begin{bmatrix} 0.5979 & 0 & 0 & 0.6106 & 0.8651 & 1.2721 \\ 0.2325 & 0.5730 & 0 & 0 & 0 & 0 \\ 0 & 0.2574 & 0.6228 & 0 & 0 & 0 \\ 0 & 0 & 0.2076 & 0.6394 & 0 & 0 \\ 0 & 0 & 0 & 0.1910 & 0.5231 & 0 \\ 0 & 0 & 0 & 0 & 0.3072 & 0.8304 \end{bmatrix}$$

produces the results in Table 5.3. The total number of cubic feet in the harvest $T = 0.1696(2.9x_0 + 7x_1 + 13x_2 + 21x_3 + 31x_4 + 43x_5)$ is also given. Equilibrium is not reached until after the year 2092. The number in each size class is given by $\hat{\mathbf{X}} = [2390, 1301, 887, 511, 205, 371]$ with harvest $T = 10278$ cubic feet. The approach to equilibrium is slow since the second largest eigenvalue of $\mathbf{B}$ is $\lambda_1 = 1$.

Table 5.3

Year	Class 0	Class 1	Class 2	Class 3	Class 4	Class 5	Total harvest
1972	2829	2714	1430	367	57	10	9 444.22
1978	1977	2213	1589	532	100	26	9 708.42
1984	1626	1728	1559	670	154	52	9 861.51
1990	1580	1368	1416	752	208	90	9 955.71
1996	1699	1151	1234	775	253	139	10 024.62
2002	1884	1055	1065	751	280	193	10 084.20
2032	2455	1249	804	512	237	376	10 278.92
2062	2410	1323	888	495	198	380	10 288.30
2092	2383	1301	893	513	204	369	10 276.35

EXERCISES

Programming exercises

The following computer programs will prove helpful in doing many of the Part A exercises that follow.

1 Modify the eigenvalue analysis program requested in Chapter 4, Exercise 2, to include the survival rate S_n for the nth age class.

2 Modify the eigenvalue analysis program of Exercise 1 to include both sexes.

Part A

3 Verify the results of Example 5.1.

4 Verify the results of Example 5.2.

5 The following information on a bobcat population in Wyoming is provided by D. Crowe [2]. The bobcat population is divided into two age classes, kittens (aged 0–1) and adults ($\geqslant$ 1 year old). Breeding takes place in March and kittens are born within two weeks of June 1. The average litter size is $2M = 2.8$ kittens/litter and sex ratios are 1–1. All females, including those born the prior spring, breed. The survival rate S_0 for kittens is strongly dependent on the density of prey. The survival rate amonst adults, which includes mortality due to trapping, is estimated to be $S_1 = 0.67$.

(a) If we census before reproduction, show that $\mathbf{X}(t+1) = \mathbf{P}\,\mathbf{X}(t)$ where

$$\mathbf{P} = \begin{bmatrix} S_0 M & S_0 M \\ S_1 & S_1 \end{bmatrix}$$

(b) Conclude that if $N(t)$ is the *total number* in the population at time t, then

$$\frac{N(t+1)}{N(t)} = S_1 + S_0 M.$$

(c) Find S_0 so that the population is stationary.
(d) In an unexploited population, the adult survival rate is 0.98 and the maximum survival rate for kittens is about 0.71. Estimate the maximum rate of increase for the population.

6 A coyote population is divided into the three age classes pup, yearling, and adult. If we census after reproduction, an appropriate Leslie matrix is (see [3])

$$\mathbf{P} = \begin{bmatrix} 0.11 & 1.5 & 1.5 \\ 0.3 & 0 & 0 \\ 0 & 0.6 & 0.6 \end{bmatrix}$$

(a) Find the dominant eigenvalue λ_0 and stable age distribution $\mathbf{S}$. In addition, find $|\lambda_1/\lambda_0|$.

(b) If $\mathbf{X}(0) = [350, 80, 151]$, project the population forward ten years. When is the stable age distribution reached to 3 correct decimal places?

7 The following information on a swan population in Chesapeake Bay is provided by Reese [4]. During a storm in March 1962, one or two mated pairs of mute swans escaped from their waterfront impoundment and founded a population which had increased to about 151 swans eleven years later. The population is divided into fledglings, 1-year-olds, 2-year-olds, and adults (aged $\geqslant 3$ years). A fledgling (i.e. a young bird just able to fly) is about four months old. Female swans breed for the first time when they are 3 years old. Almost all (96%) nests contain eggs, and the average number of fledglings per active nest is 3.1. The survival rate beyond the fledgling period is high. We will use 0.9 for this rate.

(a) If we census the population when the young swans fledge, show that

$$\mathbf{P} = \begin{bmatrix} 0 & 0 & 1.35 & 1.35 \\ 0.9 & 0 & 0 & 0 \\ 0 & 0.9 & 0 & 0 \\ 0 & 0 & 0.9 & 0.9 \end{bmatrix}$$

can serve as a Leslie matrix for the female population.
(b) Find the dominant eigenvalue λ_0 and the stable age distribution $\mathbf{S}$.
(c) Assuming the entries in matrix $\mathbf{P}$ are correct, do you think one or two mating pairs escaped and survived in 1962? Justify your answer.

8 Survival and fecundity rates for the red deer in Scotland are given in the table (adapted from [1])

Age (years)	1	2	3	4	5	6	7	8
S_k	0.907	0.987	0.992	0.990	0.987	0.992	0.953	0.942
F_k	0.000	0.000	0.266	0.282	0.338	0.380	0.383	0.391

Age (years)	9	10	11	12
S_k	0.880	0.719	0.730	0.668
F_k	0.339	0.237	0.164	0.169

(a) Find the dominant eigenvalue λ_0 and the stable age distribution **S**.

(b) Such a population has been developing virtually unexploited for hundreds of years. The 1957 estimate for the female population was 75,000. Estimate the present population.

9 Verify the results of Example 5.3.

10 K. Watt [8] gives the following population matrix for a bison population living under less favorable conditions than the population in Example 5.3.

$$\mathbf{P} = \left[\begin{array}{ccc|ccc} 0 & 0 & 0.36 & 0 & 0 & 0 \\ 0.4 & 0 & 0 & 0 & 0 & 0 \\ 0 & 0.75 & 0.9 & 0 & 0 & 0 \\ \hline 0 & 0 & 0.31 & 0 & 0 & 0 \\ 0 & 0 & 0 & 0.4 & 0 & 0 \\ 0 & 0 & 0 & 0 & 0.75 & 0.9 \end{array}\right]$$

(a) What is the fate of the population? Round off the dominant eigenvalue to two decimal places.

(b) Find $\mathbf{X}(1), \ldots, \mathbf{X}(25)$ if $\mathbf{X}(0) = [200, 120, 168, 200, 120, 168]$.

11 A forest is divided into the five size classes shown in the table and the average number of cubic feet per tree is given.

Size class	0	1	2	3	4
Girth (inches)	10–20	21–30	31–40	41–50	51–60
Cubic feet/tree	2.5	8.6	16.1	25.9	40.8

The parameters in the forest harvesting model are $c_0 = c_1 = 0$, $c_2 = 1.5$, $c_3 = 3.4$, and $c_4 = 4.8$. The transition probabilities are $b_0 = 0.31$, $b_1 = 0.41$, $b_2 = 0.38$, and $b_3 = 0.32$.

(a) If we must harvest a fixed percentage $p = 1 - h$ of each size class, find the largest possible p satisfying the constraints in the model, and find the size distribution in the forest.
(b) If $\mathbf{X}(0) = [2012, 1046, 743, 291, 340]$, find the structure of the forest and the total harvest in cubic feet over the next ten time periods if the harvesting policy in (a) is used repeatedly.
(c) Find the equilibrium structure $\hat{\mathbf{X}}$ of the forest and the resulting sustainable harvest in cubic feet.

Part B

12 Prove Theorem 5.2.

13 Prove Theorem 5.4.

14 (a) If $p_{n-1}(a) = 0$ where $a > 0$, show that if $F_n > 0$, then $p_n(a) < 0$. (*Hint*: See Theorem 4.1 for the recursion between p_n and p_{n-1}.)
(b) Use (a) to show that $q_n(\lambda)$ has a positive real zero $b > a$, where $p_{n-1}(a) = 0$.

15 In the forest harvesting model, let $\mathbf{H} = (1/\lambda)\mathbf{I}$ determine the harvesting policy corresponding to the largest eigenvalue $\lambda\,(> 1)$ of the matrix

$$(\mathbf{I} - \mathbf{C})^{-1}\ (\mathbf{A} - \mathbf{C}).$$

and let $\mathbf{P} = \mathbf{AH} + \mathbf{C}(\mathbf{I} - \mathbf{H})$.

(a) Given that $\mathbf{X}(0) = \mathbf{X}_0$ and $\mathbf{X}(t+1) = \mathbf{P}\,\mathbf{X}(t)$, prove that $\hat{\mathbf{X}} = \lim_{t\to\infty} \mathbf{X}(t)$ exists.
(b) Prove that $\hat{\mathbf{X}} = (\mathbf{AH} + \mathbf{C}[\mathbf{I} - \mathbf{H}])\,\hat{\mathbf{X}}$.
(c) Is $\hat{\mathbf{X}}$ independent of $\mathbf{X}(0)$?

16 A population would ordinarily grow according to the equation $\mathbf{X}(t+1) = \mathbf{P}\,\mathbf{X}(t)$ but migration into its territory occurs at the rate of

$$\mathbf{B} = [b_0, b_1, \ldots, b_n]$$

per unit time. Thus b_k additional members of age class k are added each time period and so

$$\mathbf{X}(t+1) = \mathbf{P}\,\mathbf{X}(t) + \mathbf{B}.$$

(a) If $(\mathbf{I} - \mathbf{P})^{-1}$ exists, show that the constant vector $\mathbf{X}_p = (\mathbf{I} - \mathbf{P})^{-1}\mathbf{B}$ is a solution.
(b) Show that if $\mathbf{X}(t)$ is a second solution, $\mathbf{X}_1(t) = \mathbf{X}(t) - \mathbf{X}_p$ satisfies $\mathbf{X}_1(t+1) = \mathbf{P}\,\mathbf{X}_1(t)$.
(c) Conclude that $\mathbf{X}(t) = \mathbf{P}^t\,\mathbf{C} + (\mathbf{I} - \mathbf{P})^{-1}\mathbf{B}$, where $\mathbf{C}$ is a constant n-vector.
(d) Find $\mathbf{C}$ in terms of $\mathbf{X}(0)$.
(e) Under what conditions does $\hat{\mathbf{X}} = \lim_{t\to\infty} \mathbf{X}(t)$ exist?

REFERENCES

[1] J. R. Beddington and D. B. Taylor, 'Optimum Age Specific Harvesting of Population', *Biometrics* (1973), **29**, 801–809.
[2] D. Crowe, 'Model for Exploited Bobcat Populations in Wyoming', *J. Wildlife Management* (1975), **39**(2); 408–415.
[3] C. H. Nellis and L. Keith, 'Population Dynamics of Coyotes in Central Alberta, 1964-68', *J. Wildlife Management* (1976), **40**(3), 389–399.
[4] J. G. Reese, 'Productivity and Management of Feral Mute Swans in Chesapeake Bay', *J. Wildlife Management* (1975), **39**(2), 280–286.
[5] M. B. Usher, 'A Matrix Model for Forest Management', *Biometrics* (1969), **25**, 309–315.
[6] M. B. Usher, *Biological Management and Conservation*, London: Chapman and Hall, 1973.
[7] M. B. Usher, 'Extensions to Models, used in Renewable Resource Management, which Incorporate an Arbitrary Structure', *J. Environmental Management*, (1976), **4**, 123–140. See in particular p. 136.
[8] K. E. F. Watt, *Ecology and Resource Management*, McGraw-Hill, 1968.

CHAPTER 6

Harvesting Matrices and Linear Programming

In this chapter we will model the influence of man on a population or renewable resource. We wish to manage the resource so that the value of the harvest is maximized subject to various conservation constraints. Which age (or size) classes should be subjected to harvesting? How much of a given age class C_k should be taken? Can a particular harvesting policy be implemented year after year with *large sustained yields*? The model presented here was originated by W. G. Doubleday [3] and has been developed and generalized by several authors.† The mathematical solution involves the optimization technique called *linear programming.*

6.1 MODELING THE HARVEST

We first assume that the population is censused after reproduction and that then the harvest is taken. The harvest is then followed by a long period of uninhibited growth. Let h_k = the percentage of class C_k *left after harvesting*, and let

$$\mathbf{H} = \begin{bmatrix} h_1 & 0 & 0 & \dots & 0 \\ 0 & h_2 & 0 & \dots & 0 \\ 0 & 0 & h_3 & \dots & 0 \\ \vdots & \vdots & \vdots & & \vdots \\ 0 & 0 & 0 & \dots & h_n \end{bmatrix}$$

† A list of recent papers is given at the end of the chapter.

As in Chapter 5, **H** is called a *harvesting matrix* and specifies the harvesting policy that will be used repeatedly.

Let $\mathbf{X}(t)$ = population structure *after harvesting*. As depicted in Fig. 6.1, the population structure after the next harvest is given by

$$\boxed{\mathbf{X}(t+1) = \mathbf{HP}\,\mathbf{X}(t)}$$

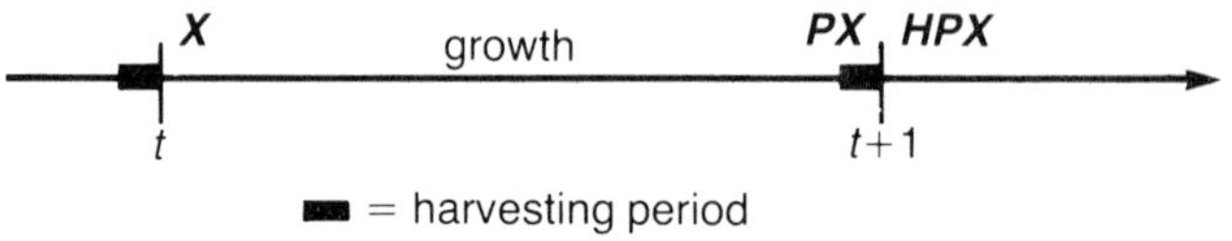

Fig. 6.1: The harvest followed by a long period of growth.

Example 6.1 What are the long range effects of the harvesting policy **H** on a population with Leslie matrix **P** where

$$\mathbf{H} = \begin{bmatrix} 1/2 & 0 & 0 \\ 0 & 1/3 & 0 \\ 0 & 0 & 1/10 \end{bmatrix} \quad \text{and} \quad \mathbf{P} = \begin{bmatrix} 0 & 9 & 12 \\ 1/3 & 0 & 0 \\ 0 & 1/2 & 0 \end{bmatrix}$$

Solution 6.1 If $\mathbf{X}(t)$ is the population structure after harvesting, then $\mathbf{X}(t+1) = \mathbf{HP}\,\mathbf{X}(t)$ and

$$\mathbf{HP} = \begin{bmatrix} 0 & 9/2 & 6 \\ 1/9 & 0 & 0 \\ 0 & 1/20 & 0 \end{bmatrix}$$

has dominant eigenvalue $\lambda_0 = 0.73834$. Thus repeated use of this harvesting policy will drive the population to extinction.

If we simply keep track of the total number T taken in the harvest at time $t+1$, then T is given by

$$T = (1-h_0)y_0 + (1-h_1)y_1 + \ldots + (1-h_n)y_n$$

where $\mathbf{Y} = \mathbf{P}\,\mathbf{X}(t)$. As usually happens, however, some age classes are more valuable than others. Let w_k = value of an individual in class C_k (measured in \$, lb., etc.). Then if $\mathbf{W} = [w_0, w_1, \ldots, w_n]$, the *total value T of the harvest at time $t+1$* is given by

$$\begin{aligned} T &= w_0(1-h_0)y_0 + w_1(1-h_1)y_1 + \ldots + w_n(1-h_n)y_n \\ &= \mathbf{W}(\mathbf{I}-\mathbf{H})\,\mathbf{Y} = \mathbf{W}(\mathbf{I}-\mathbf{H})\mathbf{P}\,\mathbf{X}(t). \end{aligned}$$

This is the function to be maximized, the so-called *objective function* in our model.

Example 6.2 For the population in Example 6.1, let $\mathbf{W} = [0, 10, 20]$. Find the values of the harvest over the next ten years given that $\mathbf{X}(0) = [90, 60, 40]$.

Solution 6.2 The value of the harvest at time $t + 1$ is

$$T = [0\ 10\ 20]\begin{bmatrix} 1/2 & 0 & 0 \\ 0 & 2/3 & 0 \\ 0 & 0 & 0.9 \end{bmatrix}\begin{bmatrix} 0 & 9 & 12 \\ 1/3 & 0 & 0 \\ 0 & 1/2 & 0 \end{bmatrix}\begin{bmatrix} x_0(t) \\ x_1(t) \\ x_2(t) \end{bmatrix}$$

$$= \frac{20}{9}x_0(t) + 9x_1(t).$$

The harvest at time $t = 1$ is then $\frac{20}{9}(90) + 9(60) = 740$. Now $\mathbf{X}(1) = \mathbf{HP\,X}(0) = [510, 10, 3]$ and the harvest at time $t = 2$ is 1223.33. Applying the formula $\mathbf{X}(t + 1) = \mathbf{HP\,X}(t)$ repeatedly leads to the results in Table 6.1, which show that T decreases rapidly beyond the second year.

Table 6.1

Time t	1	2	3	4	5	6	7	8	9	10
T	740	1223.33	650	636.33	365.78	339.83	204.10	182.11	113.38	97.86

6.2 CONSERVATION CONSTRAINTS

In order to avoid the over-exploitation of the resource (as in Example 6.2), we must impose certain constraints on the variables $h_0, h_1, \ldots, h_n$ and $x_0, x_1, \ldots, x_n$. For the constraints listed below, the notation $\mathbf{Y} \geqslant \mathbf{X}$ means that $y_k \geqslant x_k$ for $k = 0, \ldots, n$.

(1) *The population* $\mathbf{X}(t)$ *increases in all classes over the growth period, i.e.* $\mathbf{P\,X}(t) \geqslant \mathbf{X}(t)$.

(2) *After harvesting, the population is restored to its former state one time unit ago. Thus* $\mathbf{X}(t + 1) = \mathbf{X}(t)$.

We will therefore simply write $\mathbf{X}(t) = \mathbf{X}$. Let $\mathbf{Y} = \mathbf{PX}$. Then since $\mathbf{HP\,X} = \mathbf{X}$, we have $h_k y_k = x_k$ or $h_k = x_k/y_k$. *Hence once* $\mathbf{X}$ *is known* $\mathbf{H}$ *is completely determined.*

(3) $\mathbf{X} \geqslant \mathbf{0}$ *is a biologically trivial but important mathematical constraint.*

(4) *The total number in the population is constant N.* Hence $x_0 + x_1 + \ldots + x_n = N$.

We may assume, without loss of generality, that $N = 1$, for if **H** and **X** maximize T with $N = 1$, then it is easy to show that **H** and N**X** provide a solution for the general case. When $x_0 + x_1 + \ldots + x_n = 1$, we are attempting to find the *population structure* that maximizes T. T may be interpreted as the *average harvest per individual* in the population.

Constraint (4) could also take the form $\sum_{k=0}^{n} b_k x_k = c$. The constant b_k could denote the biomass of an individual in class C_k and so we could require that the total biomass be kept at a certain level.

6.3 SOLUTIONS BY LINEAR PROGRAMMING

Since $T = \mathbf{W}(\mathbf{I} - \mathbf{H})\mathbf{PX}$ and **P** and **W** will be known, T is a function of the $2n + 2$ variables $h_0, \ldots, h_n$ and $x_0, \ldots, x_n$. Some simplification, however, is possible. Since constraint (2) gives $\mathbf{HP\,X} = \mathbf{X}$,

$$T = \mathbf{WPX} - \mathbf{W}(\mathbf{HPX}) = \mathbf{WPX} - \mathbf{WX} = \mathbf{W}(\mathbf{P} - \mathbf{I})\,\mathbf{X}.$$

We must therefore maximize the *objective function* $T = \mathbf{W}(\mathbf{P} - \mathbf{I})\,\mathbf{X}$ subject to the *constraints* $(\mathbf{I} - \mathbf{P})\,\mathbf{X} \leqslant \mathbf{0}$, $\mathbf{X} \geqslant \mathbf{0}$, and $x_0 + x_1 + \ldots + x_n = 1$. This is a standard linear programming problem.

When $n = 1$ or 2, the problem can be solved graphically as illustrated in Examples 6.3 and 6.4.

Example 6.3 If $\mathbf{P} = \begin{bmatrix} 0 & 4 \\ 0.5 & 0.75 \end{bmatrix}$ and $\mathbf{W} = [1, 4]$, then we must maximize $T = \mathbf{W}(\mathbf{P} - \mathbf{I})\,\mathbf{X} = x_0 + 3x_1$ subject to the constraints $x_0 \geqslant 0, x_1 \geqslant 0, x_0 + x_1 = 1$, and $(\mathbf{I} - \mathbf{P})\,\mathbf{X} \leqslant \mathbf{0}$, i.e. $x_0 - 4x_1 \leqslant 0$ and $-\frac{1}{2}x_0 + \frac{1}{4}x_1 \leqslant 0$. The region defined by the constraints is the closed line segment K shown in Fig. 6.2. The vertices $\mathbf{A}_1$ and $\mathbf{A}_2$ are (1/3, 2/3) and (4/5, 1/5) respectively. The level curves $T = c$ are shown in Fig. 6.3. Note that the largest value of T occurs at the vertex $\mathbf{A}_1$. Since $\mathbf{P}\,\mathbf{A}_1 = [8/3, 2/3]$, $h_0 = x_0/y_0 = 1/8$ and $h_1 = x_1/y_1 = 1$. We therefore harvest 7/8 of the first class and leave the last class alone in order to maximize the value of the harvest.

The fundamental theorem asserts that the maximum value of T will always occur at a vertex of the region K defined by the constraints.

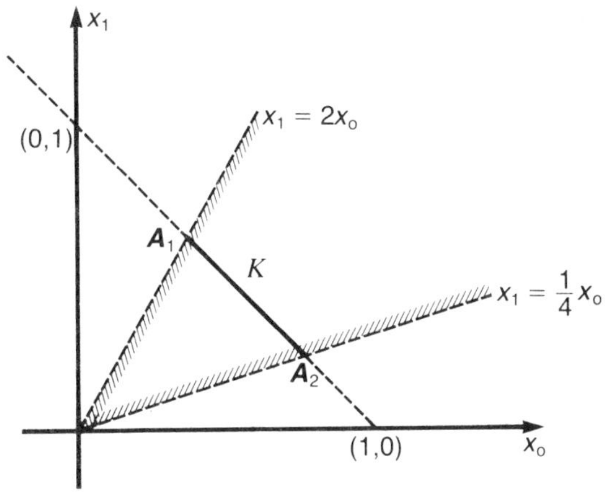

Fig. 6.2: The region defined by the constraints.

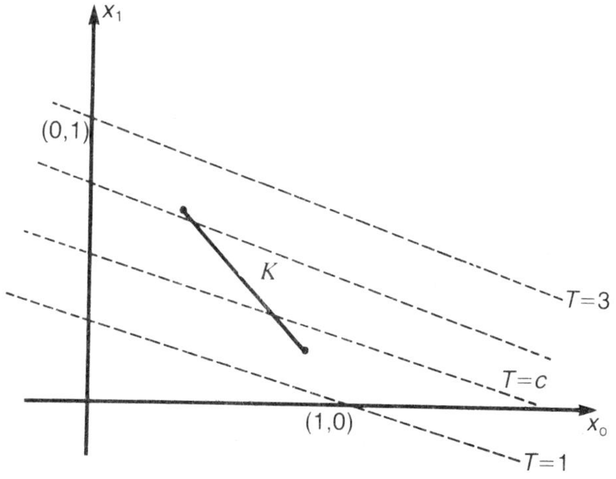

Fig. 6.3: Level curves of the objective function.

Fundamental theorem Let $T = \sum_{k=0}^{m} a_k x_k$ be a linear functional defined on $\mathbf{R}^{n+1}$, and let K be a closed and bounded convex polytope in $\mathbf{R}^{n+1}$. Then T takes on its maximum (and minimum) value at a vertex of K.

A rigorous proof of the fundamental theorem is involved, but it is not hard to see why the theorem must be true. The level curves $T = c$ define hyperplanes in $\mathbf{R}^{n+1}$ which move parallel to one another as c increases. As shown in Example 6.3 and Fig. 6.4, K will be wedged between two hyperplanes $T = c_1$ and $T = c_2$ with $c_1 < c_2$. The maximum value of T will occur at vertex $\mathbf{V}_2$. It is possible (but highly unlikely in real applications) that the hyperplane $T = c_2$ intersects K along an edge or face. Nevertheless the maximum occurs at $\mathbf{V}_2$.

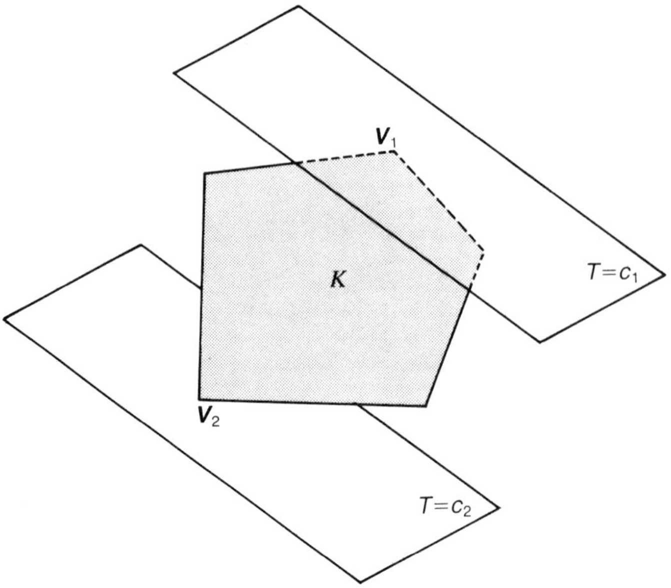

Fig. 6.4: The maximum value of T occurs at $\mathbf{V}_2$

The next example shows that once the vertices are found, we can find where the maximum value occurs by simply evaluating T at all the vertices.

Example 6.4 If **P** is the Leslie matrix in Example 6.1, then the constraints are $x_i \geqslant 0$ for $i = 0, 1, 2, x_0 + x_1 + x_2 = 1$ and the inequalities

$$x_0 - 9x_1 - 12x_2 \leqslant 0$$

$$-\frac{1}{3}x_0 + x_1 \leqslant 0$$

$$-\frac{1}{2}x_1 + x_2 \leqslant 0$$

These inequalities define the closed polygon K shown in Fig. 6.5. If $\mathbf{W} = [1, 1, 1]$, then $T = \mathbf{W}(\mathbf{P} - \mathbf{I})\mathbf{X} = -\frac{2}{3}x_0 + 8.5\,x_1 + 11x_2$. Evaluating T at each vertex (see Table 6.2), we find that T is largest at $\mathbf{A}_1$.

Now $\mathbf{Y} = \mathbf{P}\,\mathbf{A}_1 = [10/3, 2/9, 1/9]$. Thus $h_0 = x_0/y_0 = 1/5$, $h_1 = 1$, and $h_2 = 1$. We therefore harvest 4/5 of the smallest class and leave the other two classes alone.

If $\mathbf{W} = [0, 0, 1]$, then, as shown in Table 6.2, $T = \frac{1}{2}x_0 - x_2$ is largest at $\mathbf{A}_3$. Since $\mathbf{Y} = \mathbf{P}\,\mathbf{A}_3 = [9/4, 1/4, 1/8]$, we harvest 2/3 of C_0 and all of class C_2.

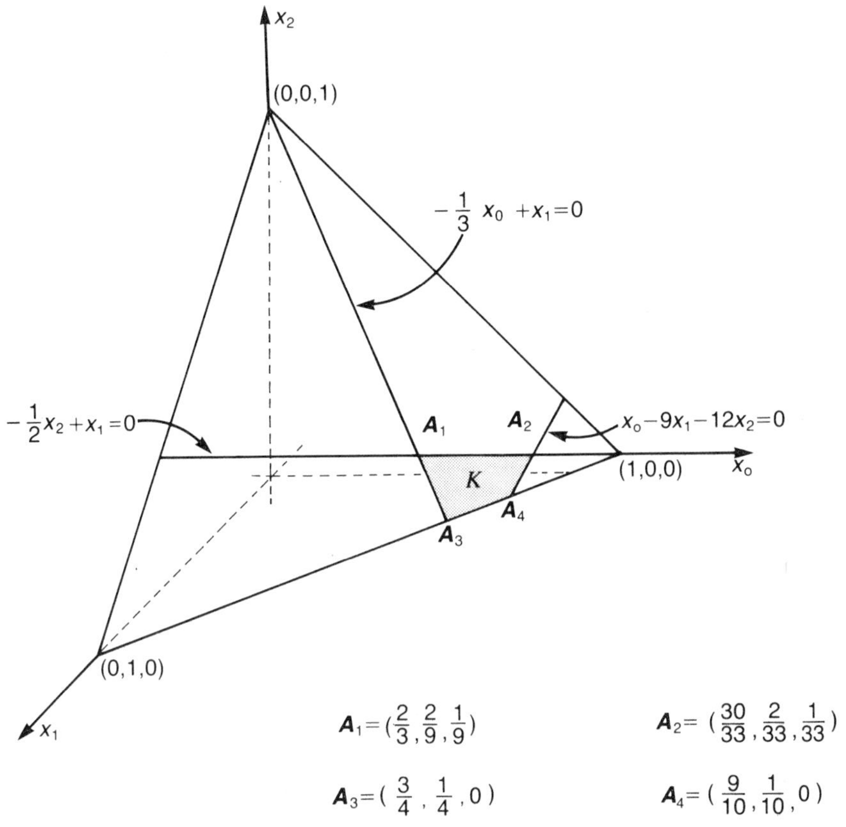

Fig. 6.5: The region defined by the constraints.

Table 6.2

W	T	at $\mathbf{A}_1$	at $\mathbf{A}_2$	at $\mathbf{A}_3$	at $\mathbf{A}_4$
[1, 1, 1]	$-\frac{2}{3}x_0 + 8.5x_1 + 11x_2$	**8/3**	8/33	13/8	1/4
[0, 0, 1]	$\frac{1}{2}x_1 - x_2$	0	0	**1/8**	1/20
[0, 1, 0]	$\frac{1}{3}x_0 - x_1$	0	**8/33**	0	1/5

If there are four or more classes, it is not possible to sketch the region K defined by the constraints as an aid in locating the vertices. The vertex where the maximum value occurs can be found algebraically by a special technique known as the *simplex method*. This method, although involving no more than matrix row operations, has many special cases and will not be described here. (See [5] for more information.) Your computer software library should have a version of the simplex method. In the IMSL library, the program is called ZX3LP. LINPRO is a popular program written in BASIC. One of the exercises will ask that you make use of a simplex method subroutine to find **X** and then find **H**. I have used the program I have written (called HARVEST) in the following examples.

Example 6.5 The Leslie matrix **P** for the female blue whale population was given in Example 5.2. Shown in Table 6.3 is the length–weight relationship for the various age classes. Except for class C_0, the formula W = 3.51 L − 192 can be used to convert lengths to weights. If **W** is the vector of weights, then **T** is the total harvest in tons. The program HARVEST produced the results in Table 6.4.

Table 6.3

Age class	0–1	2–3	4–5	6–7	8–9	10–11	⩾12
Expected length (feet)	24–68	72	78	82	85	86	87
Expected weight (long tons)	42	61	82	96	106	110	113.5

Table 6.4

Age class	Age distribution	% harvested
0–1	0.2260	0
2–3	0.1966	0
4–5	0.1710	0
6–7	0.1488	0
8–9	0.1295	0
10–11	0.1126	0
⩾ 12	0.0155	86 %

The maximum harvest is 10.711 N tons, where N is the total number of whales in the population. The optimal harvesting policy is *not* feasible. We are harvesting after reproduction and our model does not take into account the total dependence of the baby blue whale on its mother for survival. The mother must not only protect the youngster from predators but also must supply several hundred pounds of milk for a period of seven months. Hence, if we harvest the mother, we in effect harvest the baby.

Example 6.6 The two-sex Leslie matrix for the American bison was given in Example 5.3. If only adults have value for us, let **W** = [0, 0, 1, 0, 0, 1]. Table 6.5 shows the optimal harvesting solution.

Table 6.5

Age class	Age distribution	% harvested
Female calves	0.172	0
Female yearlings	0.103	0
Female adults	0.410	12.2 %
Male calves	0.197	0
Male yearlings	0.118	0
Male adults	0	100 %

The maximum harvest is 0.1455 N adults, where N is the total number in the population. This harvesting policy could possibly be implemented. Even though all adult males are taken, we could keep on hand a small number of males to assist in reproduction. Fortunately a bison is not selective in choosing its mate and so this small number of extremely fortunate adult males could be used again and again.

Suppose that we start a herd with 100 adult females, their newborn calves, and this select group of males. Then $\mathbf{X}(0) = [42, 0, 100, 48, 0, 0]$. If we apply the optimal harvesting policy **H** repeatedly, then $\mathbf{X}(t+1) = \mathbf{HP}\,\mathbf{X}(t)$ and we have the results in Table 6.6.

Table 6.6

Time t	Female calves	Female yearlings	Female adults	Male calves	Male yearlings	Male adults	Total no. harvested
0	42	0	100	48	0	0	0
1	42	25	83	48	29	0	12
2	35	25	86	40	29	0	34
3	36	21	88	41	24	0	34
4	37	22	87	42	25	0	30
5	37	22	87	42	25	0	31
6	37	22	87	42	25	0	31

Notice how quickly the optimal age distribution is reached. The *maximum sustainable yield* is 0.1455(213) = 31 adults.

6.4 MODEL MODIFICATIONS

As a generalization of the harvesting problem just presented, we can replace constraint (2) on p. 79 by:

2′. **HP** $\mathbf{X}(t) = (1 + r)\,\mathbf{X}(t)$, *where* $r > 0$.

Here, instead of restoring the population to its former state, we allow the population to grow in each class at the rate of r/unit time. We can therefore build-up the resource, but, as you would expect, the yields in the earlier years are decreased as a result. If λ_0 is the dominant eigenvalue of **P**, then this optimization problem has *no solution* if $1 + r \geqslant \lambda_0$. This should be obvious biologically since the unharvested population can grow at a maximum rate of λ_0 per time period. A formal proof can be based on the spectral decomposition theorem and is outlined in the exercise sections.

With this new constraint, constraint (1) becomes $\mathbf{P}\,\mathbf{X}(t) \geqslant (1 + r)\,\mathbf{X}(t)$ and we may write $T = \mathbf{W}(\mathbf{I} - \mathbf{H})\mathbf{PX} = \mathbf{WPX} - \mathbf{WHPX} = \mathbf{WPX} - \mathbf{W}(1 + r)\mathbf{X}$. Hence

$$\boxed{T = \mathbf{W}(\mathbf{P} - (1 + r)\mathbf{I})\mathbf{X}}$$

Example 6.7 The dominant eigenvalue for the matrix **P** in Example 6.6 is $\lambda_0 = 1.10483$ and so eventually an unharvested population could increase its numbers by 10.5% each year. We wish to design an optimal harvesting policy which will allow the population to grow by 5% each year. The program HARVEST gives the results in Table 6.7.

Table 6.7

Age class	Age distribution	% harvested
Female calves	0.1704	0
Female yearlings	0.0974	0
Female adults	0.4261	6.37 %
Male calves	0.1948	0
Male yearlings	0.1113	0
Male adults	0	100 %

The maximum harvest is 0.1139 N, where N is the total number in the population. If we apply this harvesting policy repeatedly with $\mathbf{X}(0) = [42, 0, 100, 48, 0, 0]$ we obtain the results in Table 6.8. The harvests are smaller than in Example 6.6 during the first five years, but match the sustained harvest of 31 in year six and continue to grow at the rate of 5% per year.

Table 6.8

Time t	Female calves	Female yearlings	Female adults	Male calves	Male yearlings	Male adults	Total no. harvested
0	42	0	100	48	0	0	0
1	42	25	89	48	29	0	6
2	37	25	97	43	29	0	28
3	41	22	104	47	26	0	29
4	44	25	108	50	28	0	27
5	45	26	114	52	30	0	29
6	48	27	120	55	31	0	31
7	50	29	126	58	33	0	32
8	53	30	132	60	35	0	34
9	55	32	138	63	36	0	36
10	58	33	145	66	38	0	37
11	61	35	152	70	40	0	39
12	64	37	160	73	42	0	41

Note that in Examples 6.4 to 6.6 *at most two classes were harvested*. C. Rorres [9] has shown that the optimal harvesting policy always consists of partially harvesting one class and, in some cases, completely harvesting an older class.† Such a policy is called a *two-age harvesting policy*.

As a final example, we will solve the general case of the forest harvesting model of Chapter 5.

Example 6.8 For the forest harvesting model in Chapter 5, the forest grows according to $\mathbf{X}(t+1) = (\mathbf{AH} + \mathbf{C}[\mathbf{I} - \mathbf{H}])\,\mathbf{X}(t)$ and the value of the harvest at time t is given by $T = \mathbf{W}(\mathbf{I} - \mathbf{H})\,\mathbf{X}(t)$. We wish to maximize T subject to the constraints $\mathbf{X}(t+1) = \mathbf{X}(t)$, $\mathbf{X} \geqslant \mathbf{0}$, and $\sum_{i=0}^{n} x_i = 1$. This non-standard optimization problem can be converted into the linear programming problem of this chapter.

Let $\mathbf{Z} = \mathbf{HX}$. Since $(\mathbf{AH} + \mathbf{C}[\mathbf{I} - \mathbf{H}])\,\mathbf{X} = \mathbf{X}$, we have $(\mathbf{A} - \mathbf{C})\mathbf{Z} = (\mathbf{I} - \mathbf{C})\mathbf{X}$. Now $(\mathbf{I} - \mathbf{C})^{-1} = \mathbf{I} + (1 - c_0)^{-1}\,\mathbf{C}$. Letting $\mathbf{B} = [\mathbf{I} + (1 - c_0)^{-1}\,\mathbf{C}]\,(\mathbf{A} - \mathbf{C})$, we have

$$\mathbf{X} = \mathbf{B}\,\mathbf{Z}$$

† A proof is outlined in the Part B exercises at the end of the chapter.

In terms of $\mathbf{Z}$, $T = \mathbf{WX} - \mathbf{WHX} = \mathbf{WBZ} - \mathbf{WZ} = \mathbf{W(B - I)Z}$. Note that $\mathbf{Z} \leqslant \mathbf{X}$ and if $\mathbf{B} \geqslant \mathbf{0}$, it follows that $\mathbf{BZ} \leqslant \mathbf{BX} = \mathbf{Z}$ or $(\mathbf{I} - \mathbf{B})\mathbf{Z} \leqslant \mathbf{0}$. *Hence we must maximize* $T = \mathbf{W(B - I)Z}$ *subject to the constraints*

(1) $\mathbf{Z} \geqslant \mathbf{0}$

(2) $(\mathbf{I} - \mathbf{B})\,\mathbf{Z} \leqslant \mathbf{0}$

(3) $\sum_{i=0}^{n} x_i = 1$, where $\mathbf{X} = \mathbf{BZ}$.

To illustrate the model, let $\mathbf{B}$ be the matrix resulting from the Scots pines data of Example 5.4. The program HARVEST produces the results in Table 6.9. The maximum harvest is 2.44 N cubic feet, where N is the total number of trees in the forest.

Table 6.9

Size class	Size distribution (before harvesting)	% harvested
0	0.397	64.5%
1	0.127	0
2	0.158	0
3	0.1715	0
4	0.1066	0
5	0.0395	100 %

For this harvesting policy, let $\mathbf{P} = \mathbf{AH} + \mathbf{C(I - H)}$. Applying the recursion $\mathbf{X}(t + 1) = \mathbf{P}\ \mathbf{X}(t)$ repeatedly with $\mathbf{X}(0) = [4461, 2926, 1086, 222, 27, 2]$ produces the results in Table 6.10. Equilibrium is reached slowly. The number in each size class at equilibrium is

$$\hat{\mathbf{X}} = [2941, 944, 1173, 1273, 790, 292]$$

with a maximum sustainable yield of 18 060 cubic feet. If it were possible to implement this harvesting policy, this sustainable yield would be almost two times as great as the yield in Example 5.4.

Table 6.10

Year	Class 0	Class 1	Class 2	Class 3	Class 4	Class 5	Total harvest
1972	1154	2462	1722	442	68	10	2 589.86
1978	370	1813	2055	771	145	25	1 767.50

Table 6.10 (continued)

Year	Class 0	Class 1	Class 2	Class 3	Class 4	Class 5	Total harvest
1984	282	1288	2103	1107	269	54	2 849.80
1990	477	917	1977	1378	424	100	5 192.78
1996	872	680	1767	1555	584	157	8 383.06
2002	1400	556	1536	1639	726	216	11 908.78
2008	1977	523	1324	1646	834	269	15 267.21
2032	3435	829	959	1344	921	348	21 393.05
⋮	⋮	⋮	⋮	⋮	⋮	⋮	⋮
large t	2941	944	1173	1273	790	292	18 060.47

EXERCISES

Programming exercises

1 Use a simplex method subroutine to write a program which takes the population projection matrix **P** and weighting vector **W** and outputs the optimal harvesting matrix **H** and age distribution **X**.

2 Alter the program in Exercise 1 to include the new constraint $\mathbf{HP\,X}(t) = (1 + r)\,\mathbf{X}(t)$, where $r > 0$.

Part A

In each of the following four exercises, (a) determine the long range effect of the harvesting policy **H** on a population with growth matrix **P** and (b) for the given **W**, find the values of the harvest over the next ten time periods.

3

$$\mathbf{P} = \begin{bmatrix} 0 & 2 & 5 \\ 0.5 & 0 & 0 \\ 0 & 0.6 & 0.6 \end{bmatrix} \quad \mathbf{H} = \begin{bmatrix} 1 & 0 & 0 \\ 0 & 1 & 0 \\ 0 & 0 & 0.2 \end{bmatrix} \quad \mathbf{W} = [0, 0, 5]$$

and $\mathbf{X}(0) = [200, 140, 90]$.

4

$$\mathbf{P} = \begin{bmatrix} 0 & 3 \\ 0.3 & 0.5 \end{bmatrix} \quad \mathbf{H} = \begin{bmatrix} 0.5 & 0 \\ 0 & 1 \end{bmatrix} \quad \mathbf{W} = [1, 0] \text{ and } \mathbf{X}(0) = [200, 70].$$

5

$$\mathbf{P} = \begin{bmatrix} 0 & 1 & 1 \\ 0.3 & 0 & 0 \\ 0 & 0.5 & 0.6 \end{bmatrix} \quad \mathbf{H} = \begin{bmatrix} 1 & 0 & 0 \\ 0 & 0.6 & 0 \\ 0 & 0 & 0.6 \end{bmatrix} \quad \mathbf{W} = [0, 4, 7]$$

and $\mathbf{X}(0) = [75, 30, 20]$.

6
$$\mathbf{P} = \begin{bmatrix} 0 & 2 & 4 & 6 \\ 0.2 & 0 & 0 & 0 \\ 0 & 0.4 & 0 & 0 \\ 0 & 0 & 0.6 & 0.5 \end{bmatrix} \quad \mathbf{H} = \begin{bmatrix} 1 & 0 & 0 & 0 \\ 0 & 1 & 0 & 0 \\ 0 & 0 & 0.7 & 0 \\ 0 & 0 & 0 & 0.7 \end{bmatrix}$$

$\mathbf{W} = [0.5, 1.3, 2.5, 3.7]$ and $\mathbf{X}(0) = [800, 500, 350, 100]$

For the given **P** and **W**, use the graphical method to find **X** and then **H** so that $T = \mathbf{W}(\mathbf{P} - \mathbf{I})\mathbf{X}$ is maximized.

7 $\mathbf{P} = \begin{bmatrix} 0 & 3 \\ 0.3 & 0.5 \end{bmatrix}$ and $\mathbf{W} = [1, 0]$

8 $\mathbf{P} = \begin{bmatrix} 0 & 3 \\ 0.3 & 0.5 \end{bmatrix}$ and $\mathbf{W} = [0, 1]$

9 $\mathbf{P} = \begin{bmatrix} 0 & 2 & 5 \\ 0.5 & 0 & 0 \\ 0 & 0.6 & 0.6 \end{bmatrix}$ and $\mathbf{W} = [0, 1, 1]$

10 $\mathbf{P} = \begin{bmatrix} 0 & 2 & 5 \\ 0.5 & 0 & 0 \\ 0 & 0.6 & 0.6 \end{bmatrix}$ and $\mathbf{W} = [1, 1, 0]$

11 Verify the results of Example 6.5.

12 Verify the results of Examples 6.6 and 6.7 on the American bison.

13 Let **P** be the Leslie matrix for the fish population in Exercise 4, Chapter 4. If $\mathbf{W} = [2, 8, 18, 34]$ gives the weight of a fish in grams, find the optimal harvesting policy.

14 For the bobcat population in Exercise 5, Chapter 5, let

$$\mathbf{P} = \begin{bmatrix} 1.37 & 1.37 \\ 0.6 & 0.98 \end{bmatrix}$$

and $\mathbf{W} = [0, 125]$. (In spring 1974, an adult hide was worth about \$125.)

(a) If we require $\mathbf{X}(t+1) = \mathbf{X}(t)$, find the optimal harvesting policy **H** and age structure **X**.

(b) If we require $\mathbf{X}(t+1) = 1.2\,\mathbf{X}(t)$, find the optimal harvesting policy and age structure **X**.

(c) If $\mathbf{X}(0) = [6815, 2985]$ and the harvesting policy in (b) is used repeatedly, predict population levels over the next ten years and give the value of each harvest.

15 As pointed out in Exercise 7, Chapter 5, the mute swan population in Chesapeake Bay is increasing at the rapid rate of about 40% per year. Since the swan is a burden on the aquatic vegetation in the area and is a nuisance to fishermen, some attempt should be made to limit their numbers.

(a) Design an optimal harvesting policy with $\mathbf{X}(t+1) = \mathbf{X}(t)$ which maximizes the total number of swans removed from the population.
(b) If $\mathbf{X}(0) = [28, 18, 11, 19]$, predict population levels over the next ten years if the policy in (a) is to be used repeatedly.
(c) What is $\hat{\mathbf{X}} = \lim_{t \to \infty} \mathbf{X}(t)$?

16 For the red deer population in Exercise 8, Chapter 5, the mean body weights (in kg) for the various age classes are shown in the table below.

Age class	1	2	3	4	5	6	7	8	9	10	11	⩾12
Weight	23	36	46	49	51	51	53	54	54	51	50	44

Find the optimal harvesting policy with $\mathbf{X}(t+1) = \mathbf{X}(t)$ which maximizes the total weight of the harvest.

17 For the forest in Exercise 11, Chapter 5, find the optimal harvesting policy which maximizes the total number of cubic feet in the harvest. Then, with this **H**, rework parts (b) and (c) of Exercise 11, Chapter 5.

Part B

18 Let **A** be an $n \times n$ matrix with $a_{ij} \geqslant 0$ for all i and j.

(a) If $\mathbf{X} \geqslant \mathbf{Y}$ (i.e. $x_k \geqslant y_k$ for $k = 1$ to n), prove that $\mathbf{AX} \geqslant \mathbf{AY}$.
(b) Conclude that $\mathbf{A}^n \mathbf{X} \geqslant \mathbf{A}^n \mathbf{Y}$.

19 Let $\mathbf{A} = \lambda_1 \mathbf{Z}_1 + \ldots + \lambda_n \mathbf{Z}_n$ be the spectral decomposition for **A** and assume λ_1 is the dominant eigenvalue.

(a) If $\mathbf{A} \geqslant \mathbf{0}$ and if there exists an $\mathbf{X} \geqslant \mathbf{0}$ with $\mathbf{AX} \geqslant \mathbf{X}$, then prove that $\lambda_1 \geqslant 1$.
(*Hint*: Show that $\mathbf{A}^n \mathbf{X} \geqslant \mathbf{X}$ and use Corollary 3.1.3.)
(b) If $\mathbf{A} \geqslant \mathbf{0}$ and if there exists and $\mathbf{X} \geqslant \mathbf{0}$ with $\mathbf{AX} \geqslant \alpha\, \mathbf{X}$, then prove that $\alpha \leqslant \lambda_1$.

20 Use Exercise 19 to conclude that if the linear programming program of this chapter has a solution (with $\mathbf{PX} \geqslant (1 + r)\mathbf{X}$), then $1 + r \leqslant \lambda_0$, where λ_0 is the dominant eigenvalue of $\mathbf{P}$.

21 Let $\mathbf{H}$ be the optimal harvesting matrix for the set of constraints on p. 79 to 80 and assume that $\mathbf{HP}$ has a real dominant eigenvalue ω_0.

(a) Prove that $\omega_0 = 1$.

(b) If $\mathbf{X}(t + 1) = \mathbf{HP}\,\mathbf{X}(t)$ with $\mathbf{X}(0) = \mathbf{X}_0$, then prove that

$$\hat{\mathbf{X}} = \lim_{t \to \infty} \mathbf{X}(t)$$

exists and that $\mathbf{HP}\,\hat{\mathbf{X}} = \hat{\mathbf{X}}$.

22 This exercise outlines a proof for the theorem of Rorres [9] : The optimal harvesting policy is always a 'two-age' policy, that is:

(1) Either a single class is harvested or

(2) There is a partial harvest of one class C_k with a total harvest of an older class C_j $(j > k)$.

All other classes are left alone. Let $\mathbf{R}_0, \mathbf{R}_1, \ldots, \mathbf{R}_n$ be the rows of matrix $\mathbf{P} - \mathbf{I}$ and let $\mathbf{U} = [1, 1, 1, \ldots, 1]$.

(a) Show that $y_k - x_k = \mathbf{R}_k \cdot \mathbf{X}$, the dot product of $\mathbf{R}_k$ and $\mathbf{X}$.

(b) Show that the constraint region K is defined by $\mathbf{X} \geqslant \mathbf{0}$, $\mathbf{U} \cdot \mathbf{X} = 1$, and $\mathbf{R}_k \cdot \mathbf{X} \geqslant 0$, for $k = 0, 1, \ldots, n$.

There are two types of vertices for K. An *interior vertex* has $x_i > 0$ for all i. A *boundary vertex* has $x_k = 0$ for some k (see the figure below).

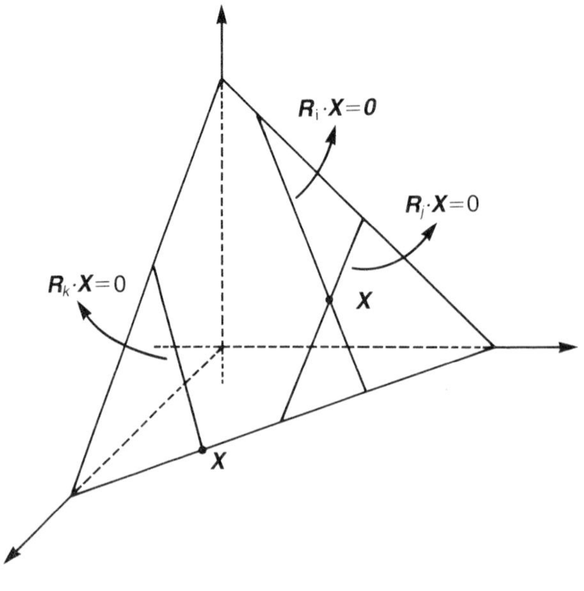

(c) Show that if $\mathbf{X}$ is an interior vertex, $\mathbf{R}_k \cdot \mathbf{X} = 0$ for n of the equations $\mathbf{R}_0 \cdot \mathbf{X} = 0, \ldots, \mathbf{R}_n \cdot \mathbf{X} = 0$.

(d) Conclude that at an interior vertex $\mathbf{X}$, $y_k = x_k$ for n of the $n + 1$ classes. Thus, if the maximum occurs at this vertex, at most one class is harvested.

(e) If $x_k = 0$ for some k, then prove that $y_{k+1} = x_{k+1} = 0$. Conclude that it is not possible to completely harvest two distinct non-empty classes.

(f) If $\mathbf{X}$ is a boundary vertex, conclude that $x_n = 0$.

(g) For $n = 2$, prove that at least one of the two classes C_0 or C_1 is left unharvested if the maximum occurs at a boundary vertex. Conclude that the theorem is true for $n = 2$.

Assume that the theorem has been established for n classes $C_0, \ldots, C_{n-1}$. If $\mathbf{X}$ is an interior vertex and the maximum occurs at $\mathbf{X}$, then part (d) shows that the theorem is valid. If the maximum occurs at a boundary vertex, then $x_n = 0$. Let $\mathbf{X}' = [x_0, \ldots, x_{n-1}]$, $\mathbf{U}' = [1, \ldots, 1]$, and $\mathbf{R}_k'$ denote row $\mathbf{R}_k$ with the last entry deleted.

(h) Show that $\mathbf{U}' \cdot \mathbf{X}' = 1$ and $\mathbf{R}_k' \cdot \mathbf{X}' \geqslant 0$ for $k = 0, \ldots, n - 1$. Complete the proof for $n + 1$ classes by using the induction hypothesis and part (e).

REFERENCES

[1] J. R. Beddington and D. B. Taylor, 'Optimum Age Specific Harvesting of a Population', *Biometrics* (1973), **29**, 801–809.

[2] J. R. Beddington, 'Age Structure, Sex Ratio and Population Density in the Harvesting of Natural Animal Populations', *J. of Applied Ecology* (1974), 913–924.

[3] W. G. Doubleday, 'Harvesting in Matrix Population Models', *Biometrics* (1975), **31**, 189–200.

[4] Gregory M. Dunkel, 'Maximum Sustainable Yields', *SIAM J. Appl. Math.* (1970), **19**, 149–164.

[5] Saul I. Gass, *Linear Programming* (4th edn), McGraw-Hill, 1975.

[6] R. Mendelssohn, 'Optimization Problems Associated with a Leslie Matrix', *American Naturalist* (1976), **110**, 339–349.

[7] William J. Reed, 'Optimum Age-specific Harvesting in a Non-linear Population Model', *Biometrics* (1980), **36**, 579–593.

[8] C. Rorres and W. Fair, 'Optimal Harvesting Policy for an Age-specific Population', *Math. Biosciences* (1975), **24**, 31–47.

[9] C. Rorres, 'Optimal Sustainable Yield of a Renewable Resource', *Biometrics* (1976), **32**, 945–948.

CHAPTER 7

Driving Functions and Non-homogeneous Linear Systems

7.1 DRIVING FUNCTIONS

Our basic compartmental model specifies how material is transferred from one compartment to another. But how does the material initially get into the system? We may suppose that the material in quickly injected into one or more compartments as would be the case when a bolus of dye is injected into the body for a particular physiological test. Mathematically, this simply corresponds to specifying an initial condition $\mathbf{X}(0) = \mathbf{X}_0$.

Alternately, we may suppose that the material is *supplied continuously* to the system from an outside source. Let $F_i(t)$ be the rate at which material enters compartment i from the outside, and let

$$\mathbf{F}(t) = [F_1(t), F_2(t), \ldots, F_n(t)].$$

Then the dynamics of the system are described by

$$\dot{\mathbf{X}} = \mathbf{A}\,\mathbf{X} + \mathbf{F}(t).$$

In this chapter, we will discuss solutions to this *non-homogeneous system.* The functions $F_1(t), \ldots, F_n(t)$ are called *driving* or *forcing functions* and arise naturally in modeling ecosystems and in tracer methods in physiology. The following are a few typical examples.

Example 7.2 Shown in Fig. 7.1 is a compartment model for the transfer of cesium-137 in an Arctic food chain. The driving function $F_1(t)$ is the fallout rate (in e.g. m Ci/km^2 per day) from the atmosphere to the vegetation at time t.

Thus $\mathbf{F}(t) = [F_1(t), 0, 0]$. Eberhardt and Hanson [1] used an exponential decay function $k\ \mathrm{e}^{-\alpha t}$ for $F_1(t)$.

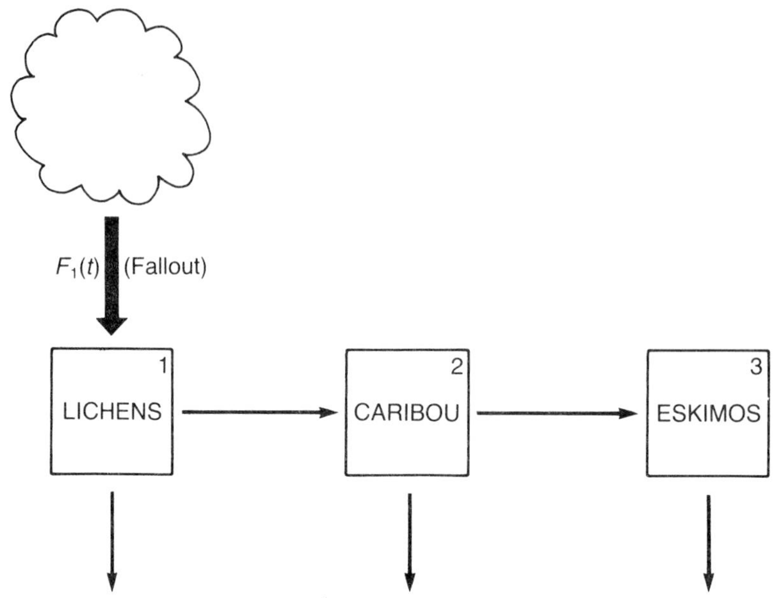

Fig. 7.1: Compartment diagram for an Arctic food chain.

Example 7.2 A dye such as indocyanine green is infused into the aorta at a fixed rate of $F_1(t) = I$ mg/min and is removed from the cardiovascular system by the liver at the unknown rate of R mg/min. Measurements of the concentration of dye are made at sampling sites S_1 and S_2 as shown in Fig. 7.2(a). With this in mind we construct the three-compartment model shown in Fig. 7.2(b). The driving function is then $\mathbf{F}(t) = [I, -R, 0]$. The object of the model is to determine the removal rate R and the flow rate through the liver given I and concentration measurements in compartments 1 and 2.

Example 7.3 A model for the flow of energy through an aquatic ecosystem is shown in Fig. 7.3. The sun drives the system supplying energy directly to compartments 1 and 2. Hence

$$\mathbf{F}(t) = [F_1(t), F_2(t), 0, \ldots, 0] .$$

We might assume that $F_1(t)$, for example, is constant or, to simulate seasonal variations, we might let $F_1(t)$ be sinusoidal in form (see Example 7.4). Although we would normally measure the mass of a compartment in kg, we must convert these measurements to an appropriate energy unit such as calories/square meter of ocean surface. Much of the energy supplied by the sun is lost to the system. Hence, a ninth compartment labelled 'losses' might be created.

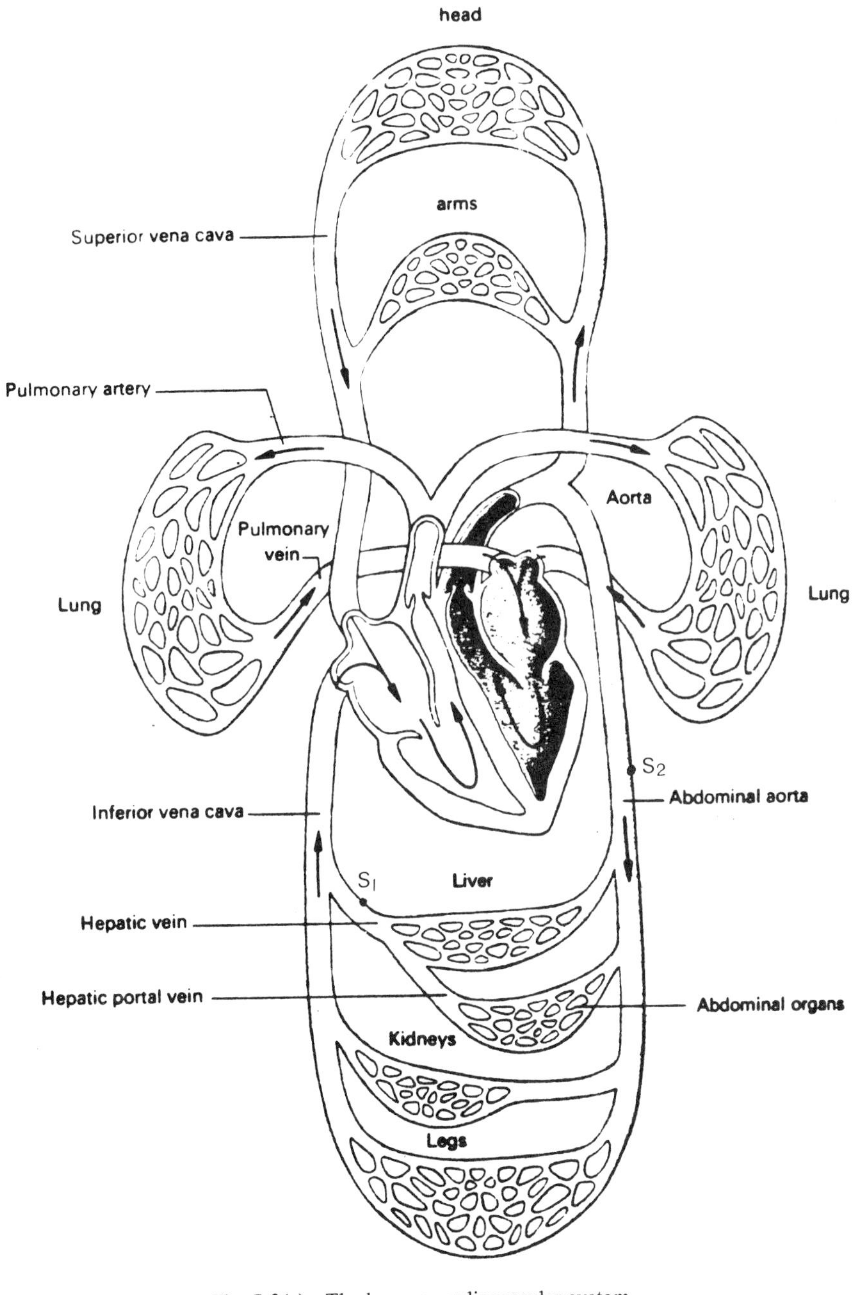

Fig. 7.2(a): The human cardiovascular system.

Adaptation of Figure 7-16 from *Living Systems: Principles and Relationships by* James M. Ford and James E. Monroe. Copyright (c) 1971 by James M. Ford and James E. Monroe. Reprinted by permission of Harper and Row, Publishers Inc.

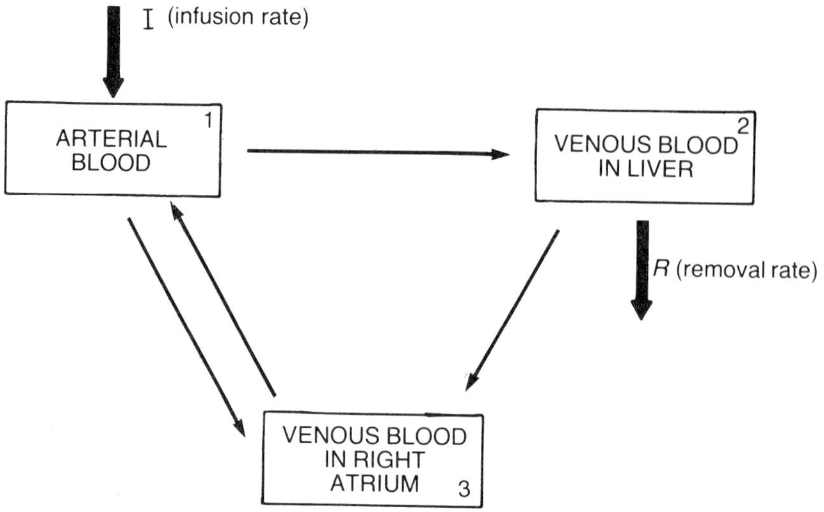

Fig. 7.2(b): Compartment diagram for the infusion of indocyanine green.

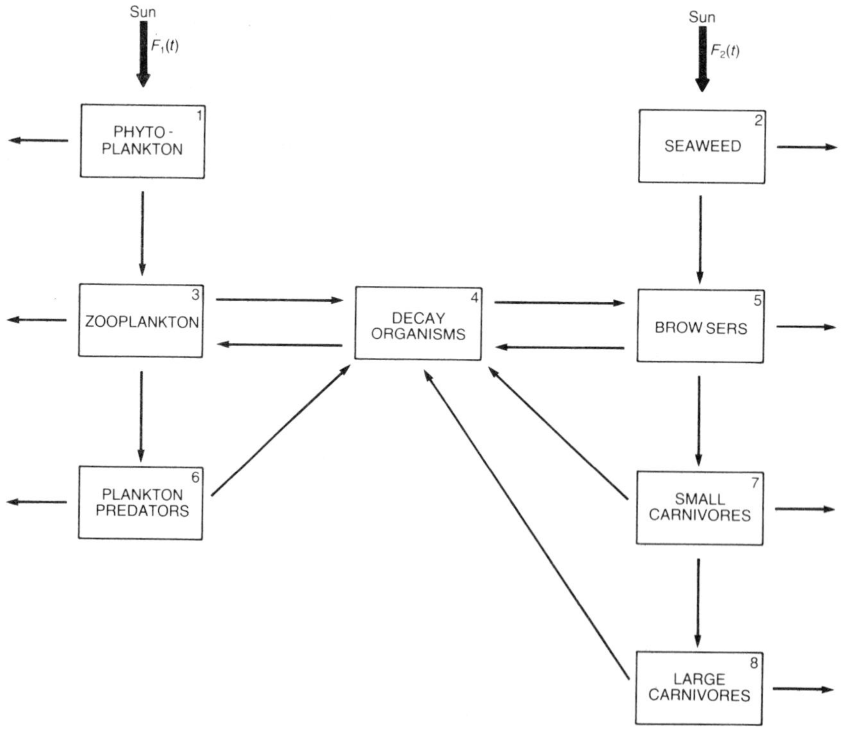

Fig. 7.3: An aquatic ecosystem and corresponding compartment diagram.

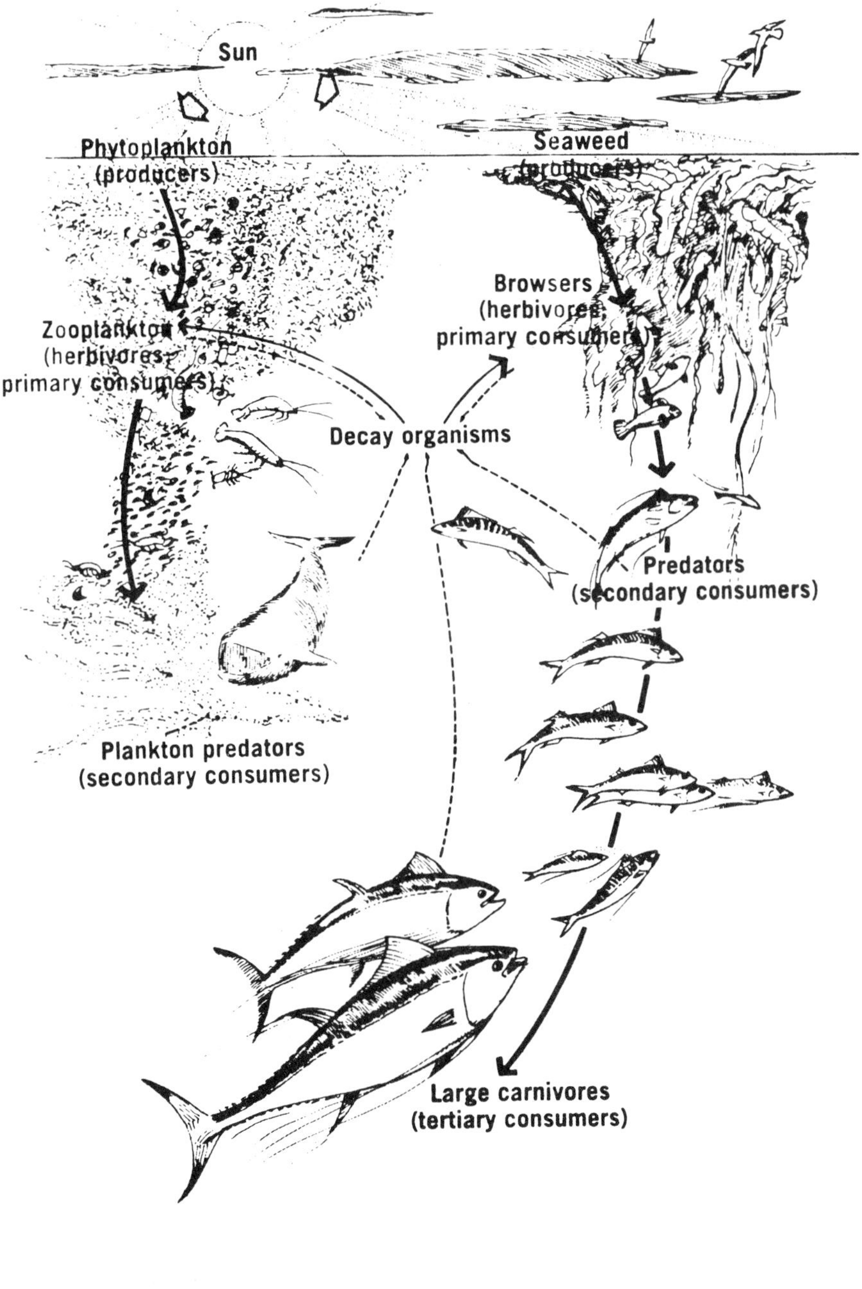

From *Ecosystems, Energy, Population* by Jonathan Turk, Janet T. Wittes, Robert Wittes and Amos Turk. Copyright (c) 1975 by W.B. Saunders Company. Reprinted by permission of CBS College Publishing.

Example 7.4 A common driving function in ecology is of the form

$$F(t) = \mu + \frac{K}{2} \sin \frac{2\pi}{p} (t - t_0)$$

where μ = average value of $F(t)$, K is the total variation of $F(t)$, p = the period, and $F(t_0) = \mu$. For example, if $F(t)$ = day length in hours,

$$F(t) = 12 + \frac{K}{2} \sin \frac{2\pi}{365} (t - 79.01721)$$

where K (which depends on latitude) is the difference between the longest and shortest day of the year, and $t_0 = 79.01721$ corresponds to 21 March, the beginning of spring. The maximum and minimum values of $F(t)$ occur around 19 June (the beginning of summer) and 19 December (the onset of winter).

7.2 SOLVING A NON-HOMOGENEOUS SYSTEM

Next let us turn to the mathematics. As we have seen in Chapter 1, the general solution to $\dot{\mathbf{X}} = \mathbf{AX}$ is $\mathbf{X}_c(t) = e^{\mathbf{A}t}\ \mathbf{C}$, where $\mathbf{C}$ is a constant n-vector. Alternately, if $\mathbf{X}_1(t), \mathbf{X}_2(t), \ldots, \mathbf{X}_n(t)$ are linearly independent solutions, the general solution may be written as

$$\mathbf{X}_c(t) = c_1 \mathbf{X}_1(t) + c_2 \mathbf{X}_2(t) + \ldots + c_n \mathbf{X}_n(t)$$

where $c_1, \ldots, c_n$ are arbitrary constants. In Chapter 2, we constructed eigenvalue–eigenvector solutions of the form $\mathbf{X}_k(t) = \mathbf{E}_k\ e^{\lambda_k t}$. Our first theorem shows that once we find a *single solution* to $\dot{\mathbf{X}} = \mathbf{A}\,\mathbf{X} + \mathbf{F}(t)$, then the general solution may be constructed.

Theorem 7.1 Suppose that $\mathbf{X}_p(t)$ is a particular solution to the differential equation $\dot{\mathbf{X}} = \mathbf{A}\,\mathbf{X} + \mathbf{F}(t)$. Then the general solution is

$$\mathbf{X}(t) = \mathbf{X}_c(t) + \mathbf{X}_p(t)$$

where $\mathbf{X}_c(t)$ is the general solution to $\dot{\mathbf{X}} = \mathbf{A}\,\mathbf{X}$.

Proof of Theorem 7.1 If $\mathbf{X}(t)$ is a second solution, let $\mathbf{Y}(t) = \mathbf{X}(t) - \mathbf{X}_p(t)$. Then

$$\begin{aligned}\dot{\mathbf{Y}}(t) &= \dot{\mathbf{X}}(t) - \dot{\mathbf{X}}_p(t) = (\mathbf{A}\,\mathbf{X}(t) + \mathbf{F}(t)) - (\mathbf{A}\,\mathbf{X}_p(t) + \mathbf{F}(t)) \\ &= \mathbf{A}(\mathbf{X}(t) - \mathbf{X}_p(t)) = \mathbf{A}\,\mathbf{Y}(t).\end{aligned}$$

Hence, $\mathbf{X}(t) = \mathbf{Y}(t) + \mathbf{X}_p(t)$ where $\mathbf{Y}(t)$ is a solution to $\dot{\mathbf{X}} = \mathbf{AX}$.

7.3 CONSTRUCTING PARTICULAR SOLUTIONS

Thus we need only find one solution to the non-homogeneous system. How then can a particular solution be constructed? Several methods are available.

In our first formula, we will obtain an integral representation of $\mathbf{X}_p(t)$ in terms of $\mathbf{A}$ and $\mathbf{F}(t)$.

When we write $\int_a^b \mathbf{X}(t)\,\mathrm{d}t$, where $\mathbf{X}(t) = [x_1(t), \ldots, x_n(t)]$, we mean

$$\left[\int_a^b x_1(t)\,\mathrm{d}t,\ \int_a^b x_2(t)\,\mathrm{d}t, \ldots,\ \int_a^b x_n(t)\,\mathrm{d}t\right]$$

In addition, we will use the following properties of the exponential matrix $\mathrm{e}^{\mathbf{A}t}$:

(1) $\dfrac{\mathrm{d}}{\mathrm{d}t}\mathrm{e}^{\mathbf{A}t} = \mathbf{A}\,\mathrm{e}^{\mathbf{A}t} = \mathrm{e}^{\mathbf{A}t}\,\mathbf{A}$

(2) $\mathrm{e}^{\mathbf{A}(t+s)} = \mathrm{e}^{\mathbf{A}t}\,\mathrm{e}^{\mathbf{A}s}$

(3) $[\mathrm{e}^{\mathbf{A}t}]^{-1} = \mathrm{e}^{-\mathbf{A}t}$

Theorem 7.2 A particular solution to $\dot{\mathbf{X}} = \mathbf{A}\,\mathbf{X} + \mathbf{F}(t)$ satisfying $\mathbf{X}_p(0) = \mathbf{0}$ is given by

$$\mathbf{X}_p(t) = \int_0^t \mathrm{e}^{\mathbf{A}(t-s)}\,\mathbf{F}(s)\,\mathrm{d}s.$$

Proof of Theorem 7.2 If $\dot{\mathbf{X}} - \mathbf{A}\,\mathbf{X} = \mathbf{F}(t)$, then $\mathrm{e}^{-\mathbf{A}t}\,(\dot{\mathbf{X}} - \mathbf{A}\,\mathbf{X}) = \mathrm{e}^{-\mathbf{A}t}\,\mathbf{F}(t)$. But by the product rule,

$$\begin{aligned}\frac{\mathrm{d}}{\mathrm{d}t}[\mathrm{e}^{-\mathbf{A}t}\,\mathbf{X}(t)] &= \mathrm{e}^{-\mathbf{A}t}\,\dot{\mathbf{X}}(t) - \mathrm{e}^{-\mathbf{A}t}\,\mathbf{A}\,\mathbf{X}(t)\\ &= \mathrm{e}^{-\mathbf{A}t}\,(\dot{\mathbf{X}}(t) - \mathbf{A}\,\mathbf{X}(t)).\end{aligned}$$

Hence $\mathrm{d}/\mathrm{d}t\,[\mathrm{e}^{-\mathbf{A}t}\,\mathbf{X}(t)] = \mathrm{e}^{-\mathbf{A}t}\,\mathbf{F}(t)$. Taking anti-derivatives, we obtain

$$\mathrm{e}^{-\mathbf{A}t}\,\mathbf{X}(t) = \int_0^t \mathrm{e}^{-\mathbf{A}s}\,\mathbf{F}(s)\,\mathrm{d}s.$$

Multiplying both sides by $\mathrm{e}^{\mathbf{A}t}$ gives the desired formula.

Corollary 7.2.1 The solution to $\dot{\mathbf{X}} = \mathbf{A}\,\mathbf{X} + \mathbf{F}(t)$ satisfying $\mathbf{X}(0) = \mathbf{X}_0$ is given by

$$\mathbf{X}(t) = \mathrm{e}^{\mathbf{A}t}\,\mathbf{X}_0 + \int_0^t \mathrm{e}^{(t-s)\mathbf{A}}\,\mathbf{F}(s)\,\mathrm{d}s$$

As elegant as the formula in Theorem 7.2 is, $\mathrm{e}^{\tau\mathbf{A}}$ must be found before the integral can be computed. When the spectral decomposition for $\mathbf{A}$ can be found, then

$$e^{\tau A} = e^{\lambda_1 \tau} Z_1 + e^{\lambda_2 \tau} Z_2 + \ldots + e^{\lambda_n \tau} Z_n$$

where $Z_1, Z_2, \ldots, Z_n$ are the spectral components of **A**. It follows that

$$\int_0^t e^{(t-s)A} F(s)\, ds = \sum_{k=1}^{n} \int_0^t e^{\lambda_k (t-s)} Z_k F(s)\, ds$$

This formula is illustrated in the following example.

Example 7.5 Find a particular solution to the differential equation

$$\dot{X} = \begin{bmatrix} 1 & -3 \\ -2 & 2 \end{bmatrix} X + \begin{bmatrix} 5e^{-2t} \\ 0 \end{bmatrix}$$

Solution 7.5 The matrix **A** has spectral decomposition

$$A = 4 \begin{bmatrix} 0.4 & -0.6 \\ -0.4 & 0.6 \end{bmatrix} + (-1) \begin{bmatrix} 0.6 & 0.6 \\ 0.4 & 0.4 \end{bmatrix}$$

Hence $e^{(t-s)A} F(s)$ is given by

$$e^{4(t-s)} \begin{bmatrix} 0.4 & -0.6 \\ -0.4 & 0.6 \end{bmatrix} \begin{bmatrix} 5e^{-2s} \\ 0 \end{bmatrix} + e^{-(t-s)} \begin{bmatrix} 0.6 & 0.6 \\ 0.4 & 0.4 \end{bmatrix} \begin{bmatrix} 5e^{-2s} \\ 0 \end{bmatrix}$$

$$= e^{4t} [2e^{-6s}, -2e^{-6s}] + e^{-t} [3e^{-s}, 2e^{-s}].$$

Thus

$$\int_0^t e^{(t-s)A} F(s)\, ds = e^{4t} \left[\frac{1}{3} - \frac{1}{3} e^{-6t}, -\frac{1}{3} + \frac{1}{3} e^{-6t} \right]$$

$$+ e^{-t} [3 - 3e^{-t}, 2 - 2e^{-t}]$$

$$= e^{4t} \left[\frac{1}{3}, -\frac{1}{3} \right] + e^{-t} [3, 2] + e^{-2t} \left[-\frac{10}{3}, -\frac{5}{3} \right]$$

This particular solution $X_p(t)$ satisfies $X_p(0) = \mathbf{0}$.

There is a far easier way of constructing a particular solution when the driving function $F(t)$ is of the form

$$F(t) = Y_1 e^{r_1 t} + Y_2 e^{r_2 t} + \ldots + Y_s e^{r_s t}$$

where $r_1, \ldots, r_s$ are either real or complex. The method is a generalization of the usual *method of undetermined coefficients* encountered in a basic course in differential equations.

Theorem 7.3 Suppose that $F(t) = Y_0 e^{rt}$, where r is *not* an eigenvalue of **A**. Then a particular solution is given by $X_p(t) = C e^{rt}$ where $C = (rI - A)^{-1} Y_0$.

Proof of Theorem 7.3 Let $\mathbf{X}_p(t) = \mathbf{C}\, e^{rt}$, where $\mathbf{C}$ is a constant vector to be determined. Then if $\mathbf{X}_p(t)$ is to satisfy $\dot{\mathbf{X}} = \mathbf{A}\,\mathbf{X} + \mathbf{Y}_0\, e^{rt}$,

$$\dot{\mathbf{X}}_p(t) = r\mathbf{C}\, e^{rt} = \mathbf{A}\,\mathbf{X}_p(t) + \mathbf{Y}_0\, e^{rt} = \mathbf{AC}\, e^{rt} + \mathbf{Y}_0\, e^{rt}$$

Hence $r\mathbf{C} = \mathbf{AC} + \mathbf{Y}_0$ or $(r\mathbf{I} - \mathbf{A})\mathbf{C} = \mathbf{Y}_0$. If r is not an eigenvalue of $\mathbf{A}$, then $\det(r\mathbf{I} - \mathbf{A}) \neq 0$ and so $(r\mathbf{I} - \mathbf{A})^{-1}$ exists. Therefore $\mathbf{C} = (r\mathbf{I} - \mathbf{A})^{-1}\, \mathbf{Y}_0$.

Example 7.6 Use the method of undetermined coefficients to find a particular solution to the differential equation in Example 7.5. Then construct the general solution.

Solution 7.6 Here $\mathbf{F}(t) = e^{-2t}\, [5, 0]$ and so $\mathbf{Y}_0 = [5, 0]$. Now

$$(-2\mathbf{I} - \mathbf{A})^{-1} = \begin{bmatrix} -3 & 3 \\ 2 & -4 \end{bmatrix}^{-1} = \begin{bmatrix} -2/3 & -1/2 \\ -1/3 & -1/2 \end{bmatrix}$$

Hence $\mathbf{C} = (-2\mathbf{I} - \mathbf{A})^{-1}\, \mathbf{Y}_0 = [-10/3, -5/3]$ and so $\mathbf{X}_p(t) = e^{-2t}\, [-10/3, -5/3]$. From the spectral decomposition given in Example 7.5, it follows that the general solution is

$$\mathbf{X}(t) = c_1\, e^{4t}\, [0.4, -0.4] + c_2\, e^{-t}\, [0.6, 0.4] + \mathbf{X}_p(t).$$

Note that when $\mathbf{F}(t) = \mathbf{Y}_0$, the particular solution is the constant vector $\mathbf{X}_p = -\mathbf{A}^{-1}\, \mathbf{Y}_0$. This is illustrated in the following example.

Example 7.7 Find a particular solution to the differential equation

$$\dot{\mathbf{X}} = \begin{bmatrix} -2 & 1 & 0 \\ 1 & -3 & 2 \\ 0 & 1 & -2 \end{bmatrix} \mathbf{X} + \begin{bmatrix} a \\ 0 \\ -b \end{bmatrix}$$

Finally, show that if $\mathbf{X}(t)$ is a solution, $\hat{\mathbf{X}} = \lim_{t\to\infty} \mathbf{X}(t)$ exists.

Solution 7.7 A particular solution is given by

$$\mathbf{X}_p = -\mathbf{A}^{-1}\, \mathbf{Y}_0 = \begin{bmatrix} 2/3 & 1/3 & 1/3 \\ 1/3 & 2/3 & 2/3 \\ 1/6 & 1/3 & 5/6 \end{bmatrix} \begin{bmatrix} a \\ 0 \\ -b \end{bmatrix}$$

$$= \frac{1}{6}\, [4a - 2b, 2a - 4b, a - 5b]$$

The eigenvalues of $\mathbf{A}$ are -4.30278, -2, and -0.697224. If $\mathbf{X}_c(t)$ denotes the general solution to $\dot{\mathbf{X}} = \mathbf{A}\,\mathbf{X}$, it follows that $\lim_{t\to\infty} \mathbf{X}_c(t) = \mathbf{0}$. Hence $\lim_{t\to\infty} \mathbf{X}(t) = \lim_{t\to\infty} (\mathbf{X}_c(t) + \mathbf{X}_p) = \mathbf{X}_p$.

Our next theorem allows us to construct a particular solution when $\mathbf{F}(t)$ contains a mixture of different types of exponential functions.

Theorem 7.4 If $\mathbf{Y}_1(t)$ is a particular solution to $\dot{\mathbf{X}} = \mathbf{A}\,\mathbf{X} + \mathbf{F}_1(t)$ and $\mathbf{Y}_2(t)$ satisfies $\dot{\mathbf{X}} = \mathbf{A}\,\mathbf{X} + \mathbf{F}_2(t)$, then $\mathbf{Y}_1(t) + \mathbf{Y}_2(t)$ is a particular solution to $\dot{\mathbf{X}} = \mathbf{A}\,\mathbf{X} + (\mathbf{F}_1(t) + \mathbf{F}_2(t))$.

This theorem is used in our next example.

Example 7.8 Find a particular solution to the differential equation

$$\dot{\mathbf{X}} = \begin{bmatrix} 1 & -3 \\ -2 & 2 \end{bmatrix} \mathbf{X} + \begin{bmatrix} 5e^{-2t} \\ -3e^{-0.5t} \end{bmatrix}$$

Solution 7.8 We will find particular solutions to

$$(1) \qquad \dot{\mathbf{X}} = \mathbf{A}\,\mathbf{X} + \begin{bmatrix} 5 \\ 0 \end{bmatrix} e^{-2t}$$

and

$$(2) \qquad \dot{\mathbf{X}} = \mathbf{A}\,\mathbf{X} + \begin{bmatrix} 0 \\ -3 \end{bmatrix} e^{-0.5t}$$

and then add the two particular solutions together. In Example 7.6, we saw that $\mathbf{Y}_1(t) = e^{-2t}\,[-10/3, -5/3]$ satisfies (1). For (2), $\mathbf{Y}_2(t) = \mathbf{C}\,e^{-0.5t}$, where

$$\mathbf{C} = (-0.5\,\mathbf{I} - \mathbf{A})^{-1}\,\mathbf{Y}_0 = \begin{bmatrix} 10/9 & 4/3 \\ 8/9 & 2/3 \end{bmatrix} \begin{bmatrix} 0 \\ -3 \end{bmatrix} = \begin{bmatrix} -4 \\ -2 \end{bmatrix}$$

A particular solution to the original differential equation is then

$$\mathbf{Y}(t) = e^{-2t}\,[-10/3, -5/3] + e^{-0.5t}\,[-4, -2].$$

If we use the identities $\cos \omega t = (e^{i\omega t} + e^{-i\omega t})/2$ and $\sin \omega t = (e^{i\omega t} - e^{-i\omega t})/2i$, then trigonometric functions can be handled using Theorems 7.3 and 7.4. One must, however, invert the complex matrices $\omega i\mathbf{I} - \mathbf{A}$ and $-\omega i\mathbf{I} - \mathbf{A}$. There is, however, a more straightforward procedure.

If $\mathbf{F}(t) = \mathbf{Y}_0 \cos \omega t + \mathbf{Z}_0 \sin \omega t$, then let $\mathbf{X}_p = \mathbf{C}_1 \cos \omega t + \mathbf{C}_2 \sin \omega t$ where $\mathbf{C}_1$ and $\mathbf{C}_2$ are constant vectors to be determined. If $\mathbf{X}_p$ is to be a solution,

$$\begin{aligned} \dot{\mathbf{X}}_p &= -\mathbf{C}_1\,\omega \sin \omega t + \mathbf{C}_2\,\omega \cos \omega t = (\mathbf{A}\mathbf{C}_1 \cos \omega t + \mathbf{A}\mathbf{C}_2 \sin \omega t) + \mathbf{F}(t) \\ &= (\mathbf{A}\mathbf{C}_1 + \mathbf{Y}_0) \cos \omega t + (\mathbf{A}\mathbf{C}_2 + \mathbf{Z}_0) \sin \omega t \end{aligned}$$

and so

$$\begin{aligned} -\mathbf{A}\mathbf{C}_1 + \omega\mathbf{C}_2 &= \mathbf{Y}_0 \\ -\omega\mathbf{C}_1 - \mathbf{A}\mathbf{C}_2 &= \mathbf{Z}_0 \end{aligned}$$

It follows that $(\omega^2\mathbf{I} + \mathbf{A}^2)\mathbf{C}_2 = \omega\mathbf{Y}_0 - \mathbf{A}\mathbf{Z}_0$ and $(\omega^2\mathbf{I} + \mathbf{A}^2)\mathbf{C}_1 = -\omega\mathbf{Z}_0 - \mathbf{A}\mathbf{Y}_0$. If $-\omega^2$ is not an eigenvalue for $\mathbf{A}^2$ (i.e. $\pm i\omega$ are not eigenvalues for $\mathbf{A}$), we may solve for $\mathbf{C}_1$ and $\mathbf{C}_2$ to obtain Theorem 7.5:

Theorem 7.5 Suppose that $\mathbf{F}(t) = \mathbf{Y}_0 \cos \omega t + \mathbf{Z}_0 \sin \omega t$ and that $-\omega^2$ is not an eigenvalue of $\mathbf{A}^2$. Then a particular solution is given by

$$\mathbf{X}_p(t) = \mathbf{C}_1 \cos \omega t + \mathbf{C}_2 \sin \omega t$$

where $\mathbf{C}_1 = (\omega^2\mathbf{I} + \mathbf{A}^2)^{-1}(-\mathbf{A}\mathbf{Y}_0 - \omega\mathbf{Z}_0)$ and $\mathbf{C}_2 = (\omega^2\mathbf{I} + \mathbf{A}^2)^{-1}(\omega\mathbf{Y}_0 - \mathbf{A}\mathbf{Z}_0)$.

Example 7.9 Construct a particular solution to the differential equation

$$\dot{\mathbf{X}} = \begin{bmatrix} 1 & -3 \\ -2 & 2 \end{bmatrix} \mathbf{X} + \begin{bmatrix} 3\cos t \\ 0 \end{bmatrix}$$

Solution 7.9 In this case we have $\mathbf{Y}_0 = [3, 0]$ and $\mathbf{Z}_0 = \mathbf{0}$. Now

$$(\omega^2\mathbf{I} + \mathbf{A}^2)^{-1} = (\mathbf{I} + \mathbf{A}^2)^{-1} = \begin{bmatrix} 8 & -9 \\ -6 & 11 \end{bmatrix}^{-1} = \begin{bmatrix} 11/34 & 9/34 \\ 6/34 & 8/34 \end{bmatrix}$$

and so

$$\mathbf{C}_1 = \begin{bmatrix} 11/34 & 9/34 \\ 6/34 & 8/34 \end{bmatrix} \begin{bmatrix} -1 & 3 \\ 2 & -2 \end{bmatrix} \begin{bmatrix} 3 \\ 0 \end{bmatrix} = \begin{bmatrix} 21/34 \\ 15/17 \end{bmatrix}$$

and

$$\mathbf{C}_2 = \begin{bmatrix} 11/34 & 9/34 \\ 6/34 & 8/34 \end{bmatrix} \begin{bmatrix} 3 \\ 0 \end{bmatrix} = \begin{bmatrix} 33/34 \\ 9/17 \end{bmatrix}$$

Therefore $\mathbf{X}_p(t) = \left[\frac{21}{34} \cos t + \frac{33}{34} \sin t,\ \frac{15}{17} \cos t + \frac{9}{17} \sin t\right]$.

EXERCISES

Programming exercise

1 Write a program that constructs a particular solution to $\dot{\mathbf{X}} = \mathbf{A}\,\mathbf{X} + \mathbf{F}(t)$ when (a) $\mathbf{F}(t) = \mathbf{Y}_0\, e^{rt}$ for r real, or (b) $\mathbf{F}(t) = \mathbf{Y}_0 \cos \omega t + \mathbf{Z}_0 \sin \omega t$.

Part A

2 If compartment (1) is a *source* in a compartment model, show that it gives rise to a forcing function

$$\mathbf{F}(t) = [F_2(t), \ldots, F_n(t)]$$

for the other $m -$ compartments, where $F_k(t) = x_1(0)\, a_{k1}\, e^{-at}$, $a = \sum_{j \neq 1} a_{j1}$.

3 Find a driving function of the form $F(t) = \mu + (K/2) \sin \omega(t - t_0)$ which fits the following facts:

(a) The period is 24 hours.
(b) $t = 0$ corresponds to midnight.
(c) The maximum value of 50 occurs at 2 p.m. and the minimum value is 0.

4 Find the general solution to the differential equation $\dot{x} = -\lambda x + I$ arising from a single compartment model with constant driving function.

5 Find the general solution to the differential equation $\dot{x} = -\lambda x + k\, e^{-\alpha t}$ arising from a single compartment model with exponential driving function.

6 Find the general solution to $\dot{x} = -\lambda x + k \sin \omega t$ arising from a single compartment model with periodic driving function $k \sin \omega t$.

Use the integral formula in Theorem 7.2 to construct a particular solution.

$$\dot{\mathbf{X}} = \begin{bmatrix} -4 & 2 \\ -2.5 & 2 \end{bmatrix} \mathbf{X} + \mathbf{F}(t)$$

where $\mathbf{F}(t)$ is given by:

7 $\begin{bmatrix} 3\, e^{-2t} \\ 0 \end{bmatrix}$ 8 $\begin{bmatrix} 4 \\ -1 \end{bmatrix}$

(*Hint*: You must first find the spectral decomposition for $\mathbf{A}$.)

Construct a particular solution to the differential equation $\dot{\mathbf{X}} = \mathbf{A}\,\mathbf{X} + \mathbf{F}(t)$ where

$$\mathbf{A} = \begin{bmatrix} 2 & 1 & 0 \\ -2 & 1 & 1 \\ 3 & 0 & -1 \end{bmatrix}$$

and $\mathbf{F}(t)$ is given by:

9 $\begin{bmatrix} 4 \\ -1 \\ 3 \end{bmatrix}$ 10 $\begin{bmatrix} 2 \\ 2 \\ -5 \end{bmatrix} e^{-t}$

11 $\begin{bmatrix} 4 \\ 2\,e^{-t} \\ e^{-3t} \end{bmatrix}$ **12** $\begin{bmatrix} \cos t \\ \sin t \\ 0 \end{bmatrix}$

13 For the compartmental model shown below, solve the differential equation

$$\dot{\mathbf{X}} = \mathbf{A}\,\mathbf{X} + [10\,e^{-0.08t}, 0, 0]$$

with $\mathbf{X}(0) = \mathbf{0}$.

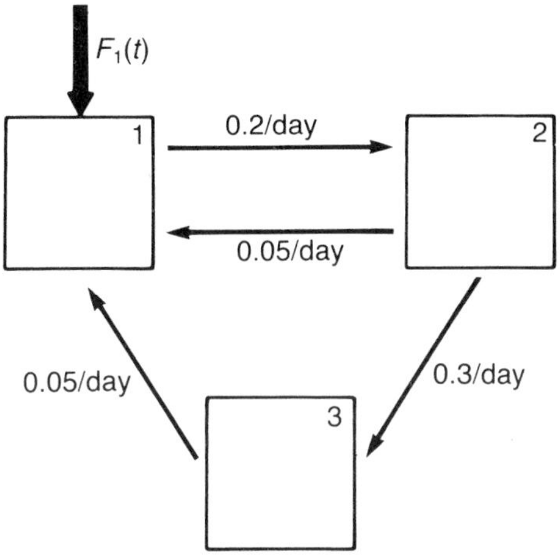

What is $\hat{\mathbf{X}} = \lim_{t \to \infty} \mathbf{X}(t)$?

14 For the compartmental model shown below, solve the differential equation

$$\dot{\mathbf{X}} = \mathbf{A}\,\mathbf{X} + [10 + 10 \sin \frac{\pi}{12}\, t, 0]$$

with $\mathbf{X}(0) = [2, 3]$.

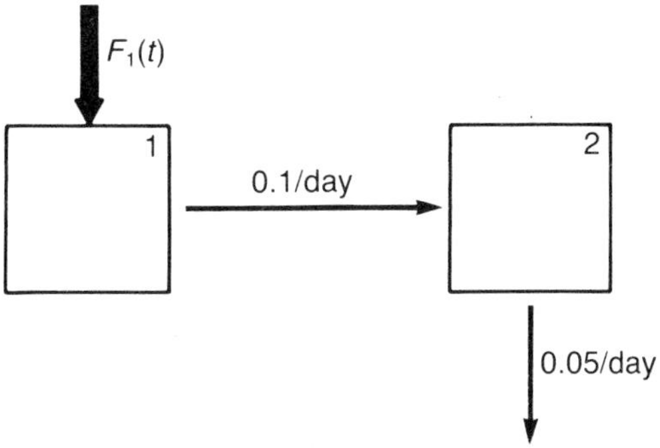

Describe the solution $\mathbf{X}(t)$ for very large t.

Part B

15 Prove Corollary 7.2.1.

16 Prove Theorem 7.4.

17 If $\dot{\mathbf{X}} = \mathbf{A}\,\mathbf{X} + (\mathbf{Y}_0 + \mathbf{Z}_0 t)$, determine conditions which guarantee that a particular solution of the form $\mathbf{X}_p(t) = \mathbf{C}_1 + \mathbf{C}_2 t$ exists.

18 If $\mathbf{A} = \sum_{i=1}^{n} \lambda_i \mathbf{Z}_i$ is the spectral decomposition for $\mathbf{A}$, determine conditions under which solutions to $\dot{\mathbf{X}} = \mathbf{A}\,\mathbf{X} + \mathbf{Y}_0\, e^{rt}$ satisfy $\lim_{t\to\infty} \mathbf{X}(t) = \mathbf{0}$.

19 If $\mathbf{A} = \sum_{i=1}^{n} \lambda_i \mathbf{Z}_i$ is the spectral decomposition for $\mathbf{A}$, determine conditions under which solutions to $\dot{\mathbf{X}} = \mathbf{A}\,\mathbf{X} + \mathbf{Y}_0$ have finite limit $\hat{\mathbf{X}}$ as $t \to +\infty$. When is $\hat{\mathbf{X}}$ independent of $\mathbf{X}(0)$?

20 Suppose that $\mathbf{A}$ has eigenvalues $\lambda_1, \lambda_2, \ldots, \lambda_n$ with $R(\lambda_k) \leqslant 0$ for all k. If $|\mathbf{F}(t)|$ is integrable on $(0, +\infty)$, show that any solution to $\dot{\mathbf{X}} = \mathbf{A}\,\mathbf{X} + \mathbf{F}(t)$ is bounded.

21 Let $\mathbf{A} = \sum_{i=1}^{n} \lambda_i \mathbf{Z}_i$ be the spectral decomposition for $\mathbf{A}$ and assume that $r = \lambda_1$ is an eigenvalue of multiplicity one.

(a) Prove that $\dot{\mathbf{X}} = \mathbf{A}\,\mathbf{X} + \mathbf{Y}_0\, e^{rt}$ has a particular solution of the form $\mathbf{X}_p(t) = (\mathbf{C}_1 + \mathbf{C}_2 t)\, e^{rt}$.

(b) Show that $\mathbf{C}_2$ can be any non-zero eigenvector corresponding to r and that $\mathbf{C}_1$ must satisfy $(r\mathbf{I} - \mathbf{A})\mathbf{C}_1 = \mathbf{Y}_0 - \mathbf{C}_2$.

22 The general formula for $e^{t\mathbf{A}}$ was developed on p. 32–33 in Chapter 2:

$$e^{t\mathbf{A}} = \sum_{i=1}^{k} e^{\lambda_i t} \left(\sum_{j=0}^{m_i - 1} \frac{t^j}{j!} \mathbf{Z}_{ij} \right)$$

Suppose that $\lambda_1 = 0$ is an eigenvalue of multiplicity m_1. Prove that the differential equation $\dot{\mathbf{X}} = \mathbf{A}\,\mathbf{X} + \mathbf{Y}_0$ has a particular solution of the form $\mathbf{X}_p(t) = \mathbf{C}_1 + \mathbf{C}_2 t + \ldots + \mathbf{C}_{m_1} t^{m_1}$.

REFERENCES

[1] L. L. Eberhardt and W. C. Hanson, 'A Simulation Model for an Arctic Food Chain', *Health Physics*, (1969), **17**, 793–806.

CHAPTER 8

An Introduction to Tracer Methods in Physiology

One of the earliest successful applications of mathematics to biology is the dye-dilution method for measuring cardiac output. This technique originated with the English physiologist George Stewart in the 1890s and was refined by the American physiologist William Hamilton in the late 1920s. In such physiological applications, a dye or radioactive tracer is injected into the bloodstream. By monitoring the concentration of tracer in various parts of the organism, we hope to obtain enough information to compute the rates at which the tracer moves through the system. Closely related is the subject of pharmacokinetics which dates back to the late 1930s and studies the rates at which drugs are distributed, metabolized, absorbed, and/or eliminated from the body.

8.1 BATH-TUB MODELS

Compartment models are often called 'bath-tub models'. Suppose that we are given n bath-tubs with pipes connecting some of the tubs. As shown in Fig. 8.1, tub (i) contains a constant volume of V_i liters and liquid flows to tub (j) from tub (i) at the rate of F_{ji} liters/minute. Next suppose that tracer (e.g. a dye or drug) is introduced into the system either through instantaneous injection or infusion via a driving function

$$\mathbf{I}(t) = [I_1(t), \ldots, I_n(t)]$$

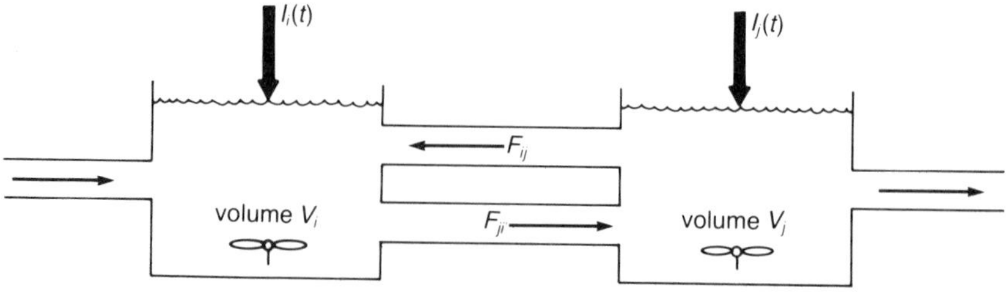

Fig. 8.1: Bath-tub Model

Let $x_i(t)$ = amount of tracer in tub (i) at time t. To insure that the concentration of tracer is uniform throughout the tub, we might imagine that blenders are placed in each tub to thoroughly mix its contents. Let $c_i(t) = x_i(t)/V_i$, the concentration in tub (i) at time t. The *flux* of tracer from tub (i) to tub (j) is then given by

$$\begin{aligned} r_{ji}(t) &= F_{ji}\,\text{(liters/minute)}\ c_i(t)\,\text{(mg/liter)} \\ &= (F_{ji}/V_i)\,x_i(t) \quad \text{(mg/minute)} \end{aligned}$$

Hence the transfer coefficient a_{ji} is given by F_{ji}/V_i. As shown in Chapters 1 and 7, the dynamics of the system are given by

$$\dot{\mathbf{X}} = \mathbf{A}\,\mathbf{X} + \mathbf{I}(t)$$

where $\mathbf{I}(t) = [I_1(t), \ldots, I_n(t)]$ specifies the rate at which the tracer enters the system from the outside and $a_{ij} = F_{ij}/V_j$. As before we can create an additional compartment (usually labelled compartment (0)) that keeps track of the tracer that leaves the system.

Shown in Fig. 8.2 is a three-tub system. In order that the volume of each tub remain constant, the flow rates into a tub must balance the flow rates out. Thus $F_{13} + F_{12} = F_{21}$, $F_{21} = F_{12} + F_{32}$, and $F_{32} = F_{13}$. In general, we have the equation

$$\boxed{\sum_{j \neq i} F_{ij} = \sum_{j \neq i} F_{ji}}$$

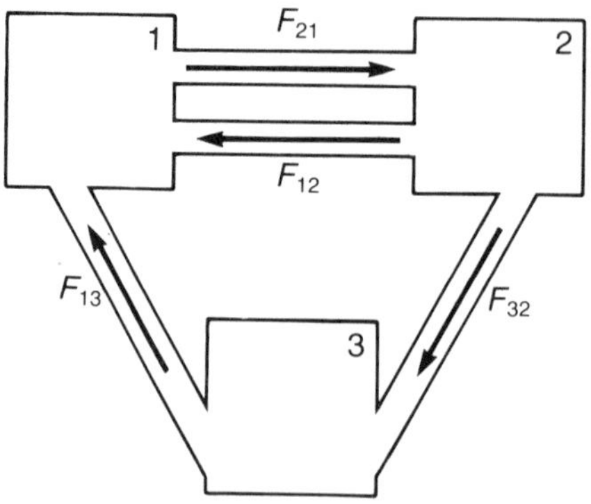

Fig. 8.2: A three-tub system

Example 8.1 Shown in Fig. 8.3 is a single compartment model with driving function $I(t)$. Hence $\dot{x} = -(F/V)x + I(t)$. It follows that $x(t) = ce^{-(F/V)t} + x_p(t)$, where x_p is a particular solution. For the case $I(t) \equiv 0$, then $x(t) = x_0 e^{-(F/V)t}$ and so the amount of tracer in the tank decays exponentially. The transfer coefficient F/V is often called the *turnover rate*.

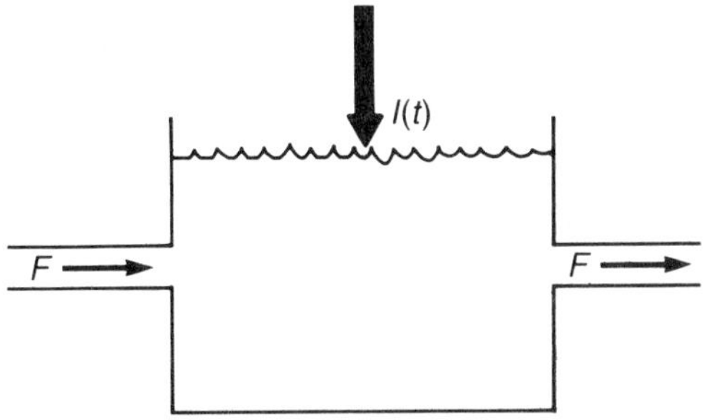

Fig. 8.3: Single compartment model with infusion

R. H. Rainey [1] has used this model to study pollution in the Great Lakes. Rainey assumes that precipitation on a given lake equals evaporation and so the flow rate in and out of the lake is constant. In addition, we must make the rather unrealistic assumption that the pollutants are uniformly distributed throughout the lake. For Lake Erie, $V \approx 458$ km^3 and $F \approx 175$ km^3/year. Hence the turnover rate is $F/V = 38.21\%$/year. If pollution were stopped completely, $I(t) \equiv 0$ and so

$$x(t) = x_0 \, e^{-0.3821\, t}$$

To estimate the amount of time it would take to clear 90% of the waste, let $x(t) = 0.1\, x_0$. This in turn gives $t \approx 6$ years. On the other hand, Lake Superior, with $V \approx 12\,221$ and $F \approx 65.2$ km^3/year, has a turnover rate 0.53%/year. It would take about 430 years to eliminate 90% of the waste.

Example 8.1 is typical of the majority of applications of compartment models to ecology. Flow rates and volumes are *measured directly* to compute transfer coefficients and make *long range predictions* about the state $\mathbf{X}(t)$ of the system. The situation in physiological applications is quite different. Flow rates and volumes are in general unknown. In fact, the whole point of the model may be to estimate a particular turnover rate, flow rate, or volume in order to judge the health of the system. Most physiological processes, however, take place over a short enough time-span that we can collect *sample concentrations*

$$c_i^*(t_1), c_i^*(t_2), \ldots, c_i^*(t_n)$$

from one or more compartments. From these partial observations of the system, we wish to estimate a_{ij} and then F_{ij} and V_j. This is the *estimation of parameters* problem. We might also want to measure the rate R at which a tracer is removed from the body by an organ. Thus some of the components $I_j(t)$ in the driving function $\mathbf{I}(t)$ may be unknown.

In the remainder of this chapter, we will present a few of the more elementary models and discuss the corresponding estimation of parameters problem. The more sophisticated two-compartment models will be presented in Chapter 9.

8.2 THE STEWART-HAMILTON METHOD FOR MEASURING CARDIAC OUTPUT

Suppose that a known quantity x_0 of dye is rapidly injected into the tank shown in Fig. 8.4. We then estimate the concentration $c(t)$ of dye in the tank at subsequent times.

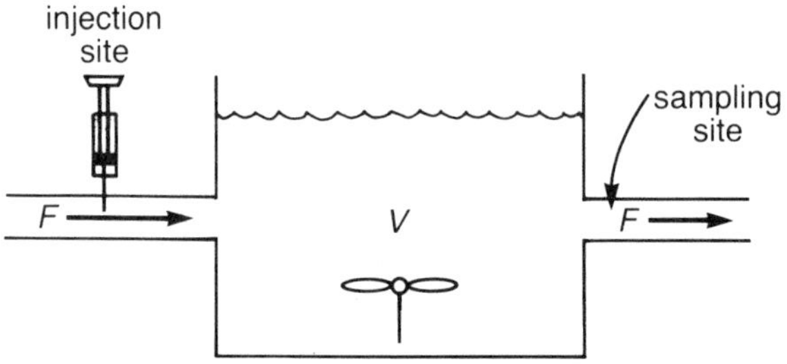

Fig. 8.4: Injection and sampling sites in a one compartment model

The rate at which dye leaves the tank is given by $r(t) = F\,c(t)$ and so the total amount of dye leaving the tank, namely x_0, is given by

$$x_0 = \int_0^{+\infty} F\,c(t)\,\mathrm{d}t \quad .$$

(Actually there will be a time T when, practically speaking, all dye is cleared from the tank.) Solving for the unknown flow rate F, we have

$$\boxed{F = \frac{x_0}{\displaystyle\int_0^{+\infty} c(t)\,\mathrm{d}t}}$$

From sample concentrations $c^*(t_1), \ldots, c^*(t_n)$, we estimate the integral using a numerical integration procedure such as the trapezoid rule or Simpson's rule.

In applying this model to measure cardiac output, the left side of the heart can serve as the 'mixing chamber' and the bolus of dye can be injected in the pulmonary artery. Alternately, the right atrium (see Fig. 7.2(a)) can serve as the mixing chamber with the dye placed in a nearby vein. The withdrawal or sampling site is a peripheral artery preferably near the aorta. A *densitometer* can make the withdrawals and dye concentration estimates as frequently as five times per second. Shown in Fig. 8.5 is a typical dye concentration curve obtained with the aid of a densitometer.

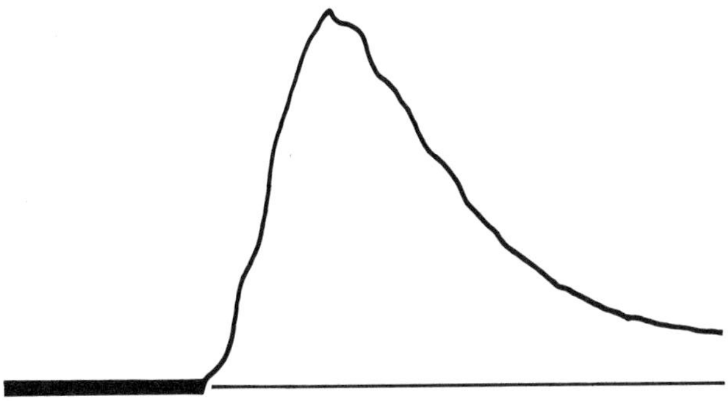

Fig. 8.5: A typical dye concentration curve

Unlike the actual cardiovascular system, our basic model assumes that once the dye has left the chamber it does not return. To minimize the effects of recirculation, it is important to select a dye that is rapidly removed from the circulatory system by an organ. In addition, because of the injection site, the dye cannot be removed or altered by the lungs. One such dye is indocyanine green (ICG) which is metabolized by the liver. Our next example gives a typical application of the method.

Example 8.2 A 5 mg bolus of ICG is injected into the right atrium. Concentration samples are then measured each second as reported in Table 8.1. Estimate the cardiac output in liters/minute.

Table 8.1

t(sec)	0	1	2	3	4	5	6	7	8	9	10	11	12
$c(t)$ (mg/l)	0	0	1.7	5.6	9.2	8.4	5.2	3.8	2.1	1.0	0.5	0.2	0

Adapted from M. R. Cullen, *Mathematics for the Biosciences*, PWS Publishers, 1983.

Solution 8.2 We are given that $x_0 = 5$ mg and $\Delta t = 1$ sec. Applying Simpson's rule we have

$$\int_0^{+\infty} c(t)\,\mathrm{d}t \approx \int_0^{12} c(t)\,\mathrm{d}t \approx \frac{1}{3}\,[0 + 4(0) + 2(1.7) + 4(5.6) + 2(9.2) + 4(8.4) + 2(5.2) + 4(3.8) + 2(2.1) + 4(1.0) + 2(0.5) + 4(0.2) + 0]$$

$$= 37.8$$

Applying the formula for F, $F = 5/37.8 = 0.1323$ liter/sec $= 7.94$ liters/min. Such a cardiac output is considered healthy.

The amount of dye in the mixing chamber is predicted to be $x(t) = x_0 e^{-(F/V)t}$ and so $c(t) = (x_0/V)\, e^{-(F/V)t}$. If we let $C = \ln c(t)$, then $C = m\,t + b$, where $m = -F/V$ and $b = \ln (x_0/V)$. Hence the relationship between $C = \ln c(t)$ and t is in theory linear. We can estimate the slope $m = -F/V$ from concentration measurements and linear regression formulas. In practice, because of the time lag in reaching the mixing chamber and incomplete mixing, the actual experimental curve (see Fig. 8.5) shows exponential decay only after a certain time t_0. We therefore perform a linear regression on C and t only after this time. As the next example shows, selecting t_0 is somewhat subjective.

Example 8.3 Using the data from Example 8.2, estimate the 'central blood volume' V.

Solution 8.3 Shown in Fig. 8.6 is the graph of $c(t)$ versus t.

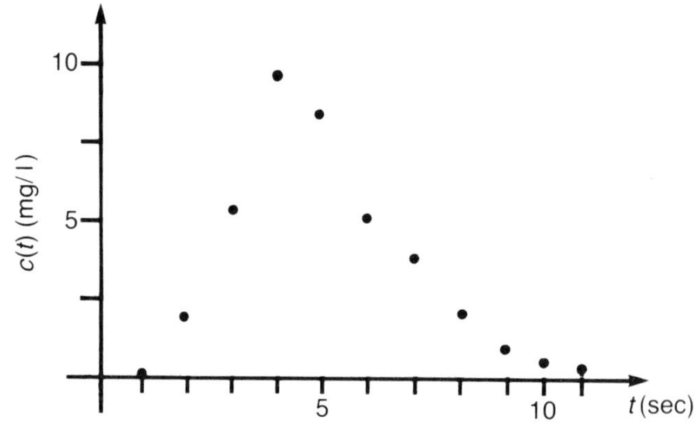

Fig. 8.6: Plot of dye concentration data

We will therefore find the regression line of $C = \ln c(t)$ versus t for $t \geqslant 4$ only. Applying the standard regression formulas gives

$$C = -0.5529\,t + 4.8683$$

with correlation coefficient $r = -0.978$. Since $F = 0.1323$ liter/sec, we have $V = -F/m \approx 0.24$ liter.

The 'central blood volume' can also be computed using the formula

$$V = \frac{F\int_0^{+\infty} t\,c(t)\,dt}{\int_0^{+\infty} c(t)\,dt}$$

Although this can be verified directly using $c(t) = (x_0/V)\,e^{-(F/V)t}$, the formula is true under more general conditions and a proof is outlined in the exercises. This formula for V is considered to be superior to the linear regression estimate. Physically, this 'central blood volume' represents volume of blood between the injection and sampling sites.

Example 8.4 Estimate V for the data in Example 8.2 using the integral formula for V given above.

Solution 8.4 We have already seen that $\int_0^{+\infty} c(t)\,dt \approx 37.8$ and $F \approx 0.1323$.

Applying Simpson's formula to $\int_0^{+\infty} t\,c(t)\,dt$ gives

$$\int_0^{+\infty} t\,c(t)\,dt \approx \int_0^{12} t\,c(t)\,dt \approx 190.933$$

Hence, from the integral formula for V, $V \approx 0.1323\,(190.933)/37.8 = 0.67$ liter.

8.3 CONTINUOUS INFUSION INTO A COMPARTMENT

As illustrated in Fig. 8.7, tracer is continuously infused into the tub at the *known rate* of I mg/min. The amount of tracer in the tub now satisfies $\dot{x} = -\lambda x + I$, where $\lambda = F/V$.

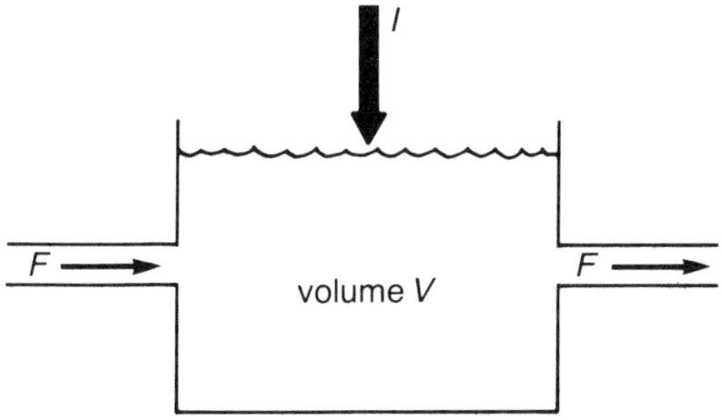

Fig. 8.7: Constant infusion into a single compartment

It follows that $x(t) = a\,e^{-\lambda t} + (I/\lambda)$ and so the concentration is given by $c(t) = a'\,e^{-\lambda t} + I/(V\lambda)$, where $a' = c(0) - I/(V\lambda)$. Hence $c_\infty = \lim_{t\to\infty} c(t) = I/(V\lambda)$ and so $a' = c(0) - c_\infty$. In summary:

$c_\infty - c(t) = (c_\infty - c(0))\,e^{-\lambda t}$
where $c_\infty = I/(V\lambda)$ and $\lambda = F/V$

An illustration of the model is provided by one of the standard glucose tolerance tests. To measure carbohydrate metabolism in a subject, glucose is infused continuously into the bloodstream at the known rate of I mg/min. Blood sugar concentrations are measured at times 0, $t_1, \ldots, t_n$ until the steady state concentration of $c_\infty = I/(V\lambda)$ mg is reached. With these measurements, the turnover rate $\lambda = F/V$ can be estimated as follows. If we let $Y = \ln\,[c_\infty - c(t)]$ then $Y = \ln\,[c_\infty - c(0)] - \lambda t$. Hence if we find the regression line $Y = mt + b$, then the turnover rate is just $-m$. The volume V can then be computed from the formula $V = I/(c_\infty \lambda)$. This is illustrated in our next example.

Example 8.5 In preparation for the glucose tolerance test, a subject fasts for two days and begins the test with glucose concentration level $c(0) = 85$ mg/deciliter. Glucose is then infused intravenously at the rate of 300 mg/min. The steady state concentration is $c_\infty = 139.2$ mg/dl, and concentration samples are shown in Table 8.2.

Table 8.2

Time t (min)	0	10	20	30	40	50
$c(t)$ (mg/dl)	85.0	105.4	120.1	127.3	131.9	134.4
$Y = \ln(c_\infty - c)$	3.9927	3.5205	2.9497	2.4765	1.9879	1.5686

The regression line of Y versus t is $Y = -0.0491\,t + 3.9773$. Hence the turnover rate is $\lambda = 0.049$/minute $= 4.9\%$/minute and $V = I/(c_\infty \lambda) = 43.89$ deciliters or about 4.4 liters. This turnover rate is in the normal range.

Fick's principle

As illustrated in Fig. 8.8, a dye of concentration c_1 flows into a tank and is removed at the constant rate of R mg/min.

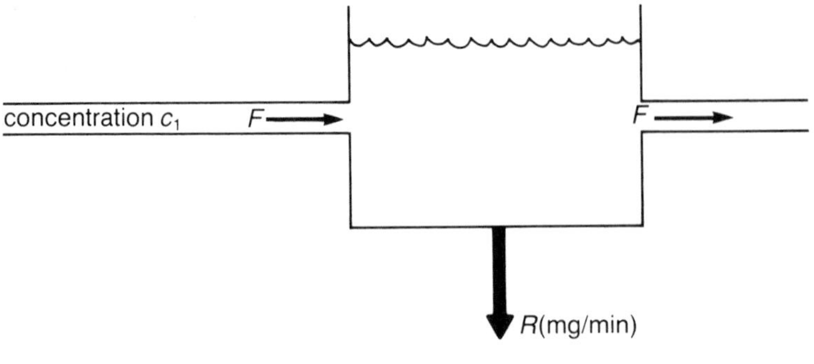

Fig. 8.8: Constant removal rate from a single compartment

If $x(t)$ is the amount of dye in the tank at time t, then

$$\dot{x} = -\frac{F}{V}x + (Fc_1 - R)$$

whose solution is of the form

$$x(t) = c\,\mathrm{e}^{-(F/V)t} + \frac{c_1F - R}{F}\,V$$

Hence $c_2 = \lim_{t\to\infty} c(t) = (c_1F - R)/F$ and so we obtain Fick's principle

$$\boxed{F = \frac{R}{c_1 - c_2}}$$

This formula can be used to measure flow rates through organs. Note, however, that the computation of F depends on our knowing the rate of removal R by the organ. How is R found? As an example, let us return to Example 7.2 in which we wish to measure hepatic blood flow. If F_1 denotes the flow rate through the liver and $F_1 + F_2$ the cardiac output, then, as shown in Fig. 8.9, we have the dynamics

$$\begin{aligned}
\dot{x}_1(t) &= -(F_1 + F_2)c_1(t) + (F_1 + F_2)c_3(t) + I \\
\dot{x}_2(t) &= F_1c_1(t) - F_1c_2(t) - R \\
\dot{x}_3(t) &= F_2c_1(t) + F_1c_2(t) - (F_1 + F_2)c_3(t)
\end{aligned}$$

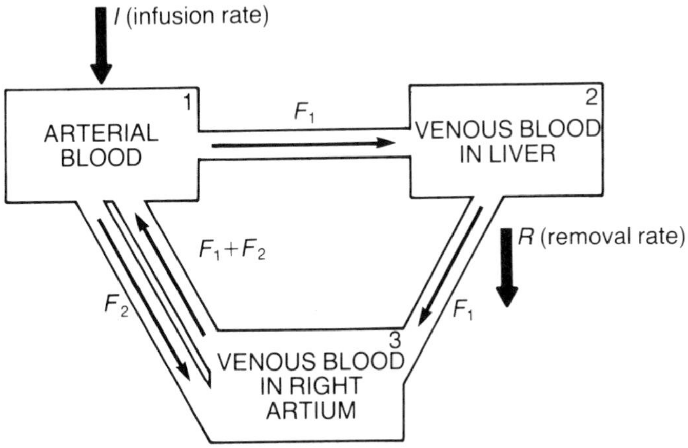

Fig. 8.9: A compartment model for hepatic blood flow

Note that we have used the formula $\sum_{j\neq i} F_{ij} = \sum_{j\neq i} F_{ji}$ in labeling the flow rates.

The transfer matrix is of the form

$$\mathbf{A} = \begin{bmatrix} -(a+b) & 0 & d \\ a & -c & 0 \\ b & c & -d \end{bmatrix}$$

and it is not hard to show then that $\lambda = 0$ is an eigenvalue of multiplicity one while the other two eigenvalues both satisfy $\text{Re}(\lambda) < 0$. Hence if $\mathbf{X}_c(t)$ is the complementary solution, $\lim_{t\to\infty} \mathbf{X}_c(t)$ exists. Therefore, the long-range solution is dependent on the form of the particular solution $\mathbf{X}_p(t)$. Since $\lambda = 0$ is an eigenvalue of multiplicity one, particular solutions are of the form $\mathbf{X}_p(t) = \mathbf{C}_1 + \mathbf{C}_2\, t$ (see Exercise 21, Chapter 7). If $\mathbf{C}_2 \neq 0$, no steady state solution will exist. We therefore need to determine conditions which will insure that $\hat{\mathbf{X}} = \lim_{t\to\infty} \mathbf{X}(t)$ exists. If $\hat{\mathbf{X}} = [\hat{x}_1, \hat{x}_2, \hat{x}_3]$ is a constant solution, let $\hat{c}_1 = \hat{x}_1/V_1$, $\hat{c}_2 = \hat{x}_2/V_2$, and $\hat{c}_3 = \hat{x}_3/V_3$. It follows that

$$\begin{aligned} 0 &= -(F_1 + F_2)\hat{c}_1 + (F_1 + F_2)\hat{c}_3 + I \\ 0 &= F_1\, \hat{c}_1 - F_1\, \hat{c}_2 - R \\ 0 &= F_2\, \hat{c}_1 + F_1\, \hat{c}_2 - (F_1 + F_2)\hat{c}_3 \end{aligned}$$

Hence $I = (F_1 + F_2)(\hat{c}_1 - \hat{c}_3)$, $R = F_1(\hat{c}_1 - \hat{c}_2)$ and $(F_1 + F_2)\hat{c}_3 = F_2\, \hat{c}_1 + F_1\, \hat{c}_2$. It follows that $I = (F_1 + F_2)\hat{c}_1 - F_2\, \hat{c}_1 - F_1\, \hat{c}_2 = F_1(\hat{c}_1 - \hat{c}_2)$, and so $I = R$. Hence we may conclude that *$I = R$ is a necessary condition for $\hat{\mathbf{X}}$ = limit $\mathbf{X}(t)$ to exist.* Conversely, when $I = R$, it is not hard to show that a constant particular solution can be constructed (see Exercise 17).

When $I = R$, the hepatic blood flow rate can be computed from the Fick's formula

$$\boxed{F_1 = \frac{I}{\hat{c}_1 - \hat{c}_2}}$$

When $I \neq R$, the long range solution will be of the form $\mathbf{C}_1 + \mathbf{C}_2 t$, with $\mathbf{C}_2 \neq 0$. The long-range concentration levels will then change at a steady rate.

To implement the model, we use ICG which is cleared from the plasma, exclusively by the liver and converted into bile. A catheter is inserted into a vein in the arm. After an initial load dose of 12 mg, dye is infused at a constant rate of I = 0.5 mg/min. For a normally functioning liver, equilibrium should be reached in about 20 minutes. Minor adjustments in I may be needed to insure equilibrium has been reached. Note that if $I \neq R$, we should notice steady increases or decreases in concentration levels. Finally, we measure concentrations from blood samples taken from an artery and a hepatic vein. For example, if $\hat{c}_1$ (arterial blood concentration) = 0.7 mg/liter and $\hat{c}_2$ (hepatic blood concentration) = 0.05 mg/liter, then F_1 = (0.5 mg/min)/(0.7 − 0.05) mg/l = 0.77 liter/min.

8.4 ELEMENTARY PHARMACOKINETICS

Compartment models in pharmacology have proven extremely useful and surprisingly accurate in predicting drug concentration levels in organs and tissues and in estimating rates at which the drug is eliminated from the body. Important clinical applications have resulted from such models. We close this chapter with two basic models.

Example 8.6 A drug is quickly injected into the bloodstream and rapidly distributes itself throughout the body. We will assume that this drug does not collect in organs and tissues and is eliminated through urine only.

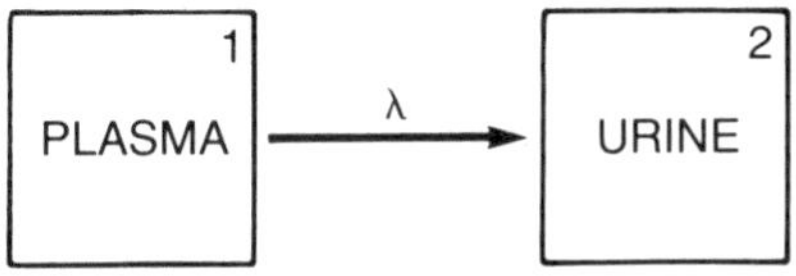

Fig. 8.10: Drug elimination through urine

From Fig. 8.10, $x_1(t) = x_0\,e^{-\lambda t}$ and the total amount of drug collected in the urine $x_2(t)$ satisfies

$$\dot{x}_2(t) = \lambda x_1(t) = \lambda x_0\,e^{-\lambda t}$$

Hence, assuming $x_2(0) = 0$, we have

$$x_2(t) = x_0(1 - e^{-\lambda t})$$

Letting $X = \ln(x_0 - x_2(t))$, we have $X = \ln x_0 - \lambda t$. Hence, given urinary excretion data $x_2^*(t_1), \ldots, x_2^*(t_n)$, we can find the regression line of X versus t and estimate the turnover rate λ. This technique is called the *sigma-minus method.*

Example 8.7 (First Order Absorption) A fixed dose of x_0 mg of drug is quickly injected into the bloodstream and makes its way to an organ where the drug is gradually destroyed or transformed. One such example is radioactive iodine-131 and the thyroid gland (Fig. 8.11).

The drug leaves the bloodstream at the rate $I(t) = x_0\lambda_1 e^{-\lambda_1 t}$. This function serves as a driving function for the second compartment. Hence

$$\dot{x}_2 = -\lambda_2 x_2 + I(t)$$

and when $\lambda_1 \neq \lambda_2$ a particular solution is given by

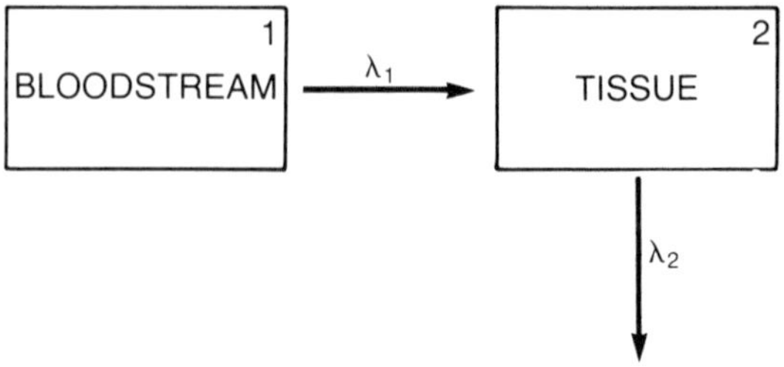

Fig. 8.11: First order absorption

$$x_p(t) = \frac{x_0\lambda_1}{\lambda_2 - \lambda_1} \mathrm{e}^{-\lambda_1 t}$$

Assuming $x_2(0) = 0$, the final solution is

$$x_2(t) = \frac{-x_0\lambda_1}{\lambda_2 - \lambda_1} [\mathrm{e}^{-\lambda_2 t} - \mathrm{e}^{-\lambda_1 t}]$$

In many cases too high a drug concentration level in the tissue can be extremely dangerous. For example, high concentrations of iodine-131 in the thyroid can produce cancerous tumors and destroy thyroid tissue. It is important then to find the maximum value of $x_2(t)$. Setting $\dot{x}_2(t) = 0$ in the expression for x_2, we can see that the maximum will occur at time

$$T = \frac{1}{\lambda_2 - \lambda_1} \ln(\lambda_2/\lambda_1)$$

and the maximum value is $x_0(\lambda_1/\lambda_2)\,\mathrm{e}^{-\lambda_1 T}$. This may be seen by setting $\dot{x}_2 = 0$ in the differential equation $\dot{x}_2 = -\lambda_2 x_2 + I(t)$.

For most drugs, λ_1 is significantly larger than λ_2. This has the important consequence that for large t

$$c_2(t) = \frac{x_2(t)}{V_2} \approx \frac{-x_0\lambda_1}{V_2(\lambda_2 - \lambda_1)} \mathrm{e}^{-\lambda_2 t}$$

and so $C_2 = \ln c_2(t) \approx -\lambda_2 t + b$ for some constant b. Hence the key turnover rate can be found by collecting dye concentration data $c_2^*(t_1)$, $c_2^*(t_2)$, . . . , $c_2^*(t_n)$ from the tissue and by finding the regression line of $C_2 = \ln c_2(t)$ versus t for large t. Thus the curve fitting technique, called the *method of exponential peeling*, will be covered in detail in Chapter 9.

Suppose that our estimate for λ_2 is 0.04 and λ_1, estimated from plasma concentration samples, is about 0.10 per hour. The maximum amount of drug in the tissue occurs at time $T = 15.27$ hours. If the maximum amount of drug in the tissue should be no more than 20 mg, then the load dose x_0 must satisfy

$$x_0 \leqslant 20\,(\lambda_2/\lambda_1)\,\mathrm{e}^{\lambda_1 T} = 36.8 \text{ mg}$$

EXERCISES

Part A

In the standard *oral glucose tolerance test*, the patient swallows a load dose of glucose (typically 1 gm/kg weight). Blood samples are then taken from a finger after 30, 45, 60, 90, 120, and 180 minutes. This test is commonly used to diagnose liver disease or diabetes. Assuming that the one compartment model on p. 111 is appropriate, estimate the turnover rate λ given the data below. Use the *excess* glucose concentrations levels.

1

t(min)	0	30	45	60	90	120	180
Concentration (mg/100 ml)	87.9	147.3	143.6	127.1	107.4	91.3	76.6

2 (Patient with cirrhosis of the liver)

t(min)	0	30	45	60	90	120	180
Concentration (mg/100 ml)	92.0	166.1	181.6	181.7	159.3	120.5	77.5

3 The form of the data in Exercises 1 and 2 suggests that a two-compartment model might be more appropriate to describe glucose metabolism following an oral dose. As illustrated below, the additional compartment is the gastro-intestinal tract.

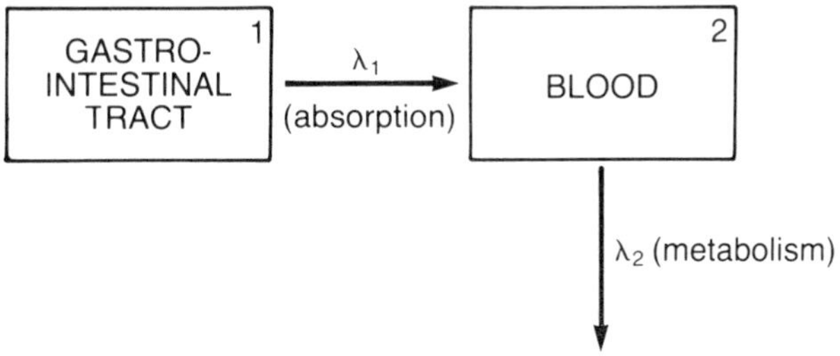

If $x_2(t)$ = amount of glucose *above the fasting level*, set up and solve the appropriate system of differential equations.

4 In the *intravenous glucose tolerance test*, a load dose of around 30 gm of glucose is injected directly into the bloodstream and blood samples are taken from an arm every ½ hour for the next two or three hours. The test is considered normal when the blood sugar level returns to the level of 100 mg/100 ml within 120 minutes. If the initial glucose concentration after the load dose is given is 300 mg/100 ml, estimate the range of normal turnover rates (in %/min).

5 The volume and net flow rate for Lake Ontario are estimated to be 1636 km^3 and 210 km^3/year. Assuming the model in Example 8.1 is appropriate, estimate the amount of time it will take to naturally displace 90% of the pollutants if pollution is stopped completely.

6 Often, in administering a load dose of drug or tracer intravenously, for safety reasons we infuse the drug into the bloodstream over a short time interval $[0, \tau]$ rather than injecting the drug all at once. For example, for the test described in Exercise 4, $\tau \approx 4$ minutes. Define

$$I(t) = \begin{cases} I, & 0 \leqslant t \leqslant \tau \\ 0, & t > \tau \end{cases}$$

(a) If the total amount of drug to be administered is x_0 mg, show that $I\tau = x_0$.

(b) Assuming that a one-compartment model is appropriate, solve $\dot{x} = -\lambda x + I(t)$ with $x(0) = 0$.

(c) Sketch the graph of the solution in (b). What happens as $\tau \to 0$?

7 In a standard 'soaking experiment' for the penetration of dye into cells, cells are placed in a large medium in which the dye concentration is constant c_0. Assuming that the rate at which the dye enters the cells is directly proportional to the concentration gradient $c_0 - c$, solve the resulting differential equation and sketch the solution.

8 When the concentration of alcohol in the bloodstream and tissues is moderate, only a small percentage of alcohol is eliminated through the lungs and kidneys. The vast majority (over 95%) of elimination takes place in the liver. It is well known that blood alcohol concentration decreases at a *constant rate* ($\approx$ 7 gm/hour). Ignoring losses through the lungs and kidneys, formulate and solve an appropriate one-compartment model.

9 Shown below is a compartmental model appropriate for the fixation of colloidal radioactive gold (^{189}Au) in the liver. If $x_1(0) = x_0$, solve the resulting system and compute $\lim_{t \to \infty} x_2(t)$. Can λ_1 and λ_2 be estimated from blood concentration measurements alone?

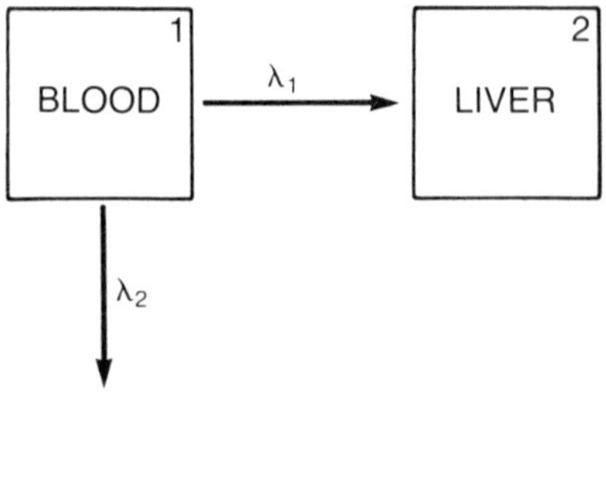

10 A drug is infused into the bloodstream at the known rate of $I = 85$ mg/min. The drug collects unmetabolized in the urine as illustrated below.

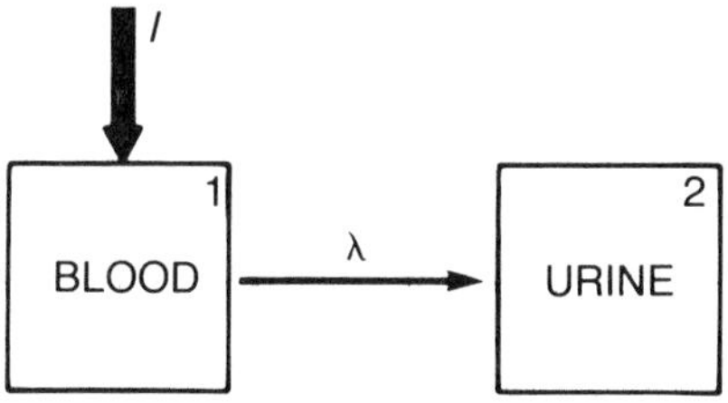

Urinary *excretion rates* are given in the table below. With this information, estimate the turnover rate λ and half-life in the bloodstream.

t (hours)	0	0.5	1.0	1.5	2.0	2.5	3.0	4.0
Excretion rate (mg/min)	0	25	40	41	46	52.7	56.8	58.0

11 Assume that when a single dose of x_0 mg of drug is taken, the amount in the bloodstream after t minutes is given by

$$x(t) = x_0 \, e^{-\lambda t}$$

If individual doses of x_0 mg are given every τ minutes,

(a) Show that $x(n\tau) = x_0(1 - a^{m+1})/(1 - a)$, where $a = e^{-\lambda\tau}$.
(b) Show that the maximum value of $x(t)$ is $x_0/(1 - e^{-\lambda\tau})$.
(c) The maximum safe concentration in the bloodstream is 15 mg/1 and the volume of distribution is 10 liters. If the half-life of the drug is 6 hours and individual doses are 20 mg, how frequently can the drug be safely administered?

A 5 mg bolus of the dye indocyanine green is injected into the right atrium. Use the concentration measurements given below to estimate the cardiac output F and central blood volume V.

12

t (sec)	0	2	4	6	8	10	12	14	16	18	20
$c(t)$ (mg/l)	0	1.6	3.7	6.8	5.2	3.1	2.4	1.8	0.6	0.2	0

13

t (sec)	0	2	4	6	8	10	12	14	16	18	20	22	24
$c(t)$ (mg/l)	0	3.2	6.4	7.0	7.2	6.8	5.4	3.1	2.6	2.0	1.2	0.4	0

14 For the matrix **A** on p. 118, show that the two non-zero eigenvalues both have negative real part. Conclude that if $\mathbf{X}(t)$ is a solution to $\dot{\mathbf{X}} = \mathbf{A}\,\mathbf{X}$, then $\lim_{t\to\infty} \mathbf{X}(t)$ exists.

15 For the model illustrated on p. 117, show that when $I = R$ a constant particular solution can be constructed.

16 Verify that the maximum value of $x_2(t)$ in Example 8.7 is

$$x_0(\lambda_1/\lambda_2)\,e^{-\lambda_1 T}$$

where $T = (\lambda_2 - \lambda_1)^{-1} \ln(\lambda_2/\lambda_1)$.

17 If I = rate of oxygen consumption (as measured by a respiratory spirometer), what concentration measurements are needed in order to use Fick's principle to compute cardiac output F?

Part B

18 As illustrated below, dye is infused into a single compartment at the rate of $I(t)$ mg/min for $0 \leqslant t \leqslant \tau$, and the flow rate through the tank is $F(t)$ liters/min.

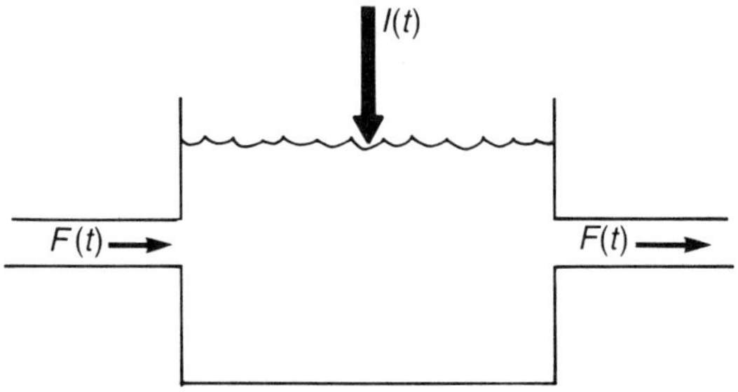

(a) Show that $\int_0^{\tau} I(t)\,dt = \int_0^{+\infty} F(t)c(t)\,dt$.

(b) Show that the proportion of dye that leaves by time t is

$$\frac{1}{x_0}\int_0^{t} F(s)c(s)\,ds$$

where $x_0 = \int_0^{\tau} I(t)\,dt$.

(c) Let T be the random variable 'length of time a dye molecule stays in the tank'. Show that the *mean transit time* is given by

$$\bar{t} = \frac{\int_0^{+\infty} tF(t)c(t)\,dt}{\int_0^{+\infty} F(t)c(t)\,dt}$$

(d) When the flow rate is constant, show directly that the mean transit time is just V/F. Next deduce the formula for V given on p. 114.

REFERENCES

[1] R. H. Rainey, 'Natural Displacements of Pollution from the Great Lakes', *Science* (1969), **155**, 1242–1243.

FURTHER READING

D. A. Bloomfield (ed.), *Dye Curves: Theory and Practice of Indicator Dilution*, Univ. Park Press (Baltimore), 1974.

J. A. Jacquez, *Compartmental Analysis in Biology and Medicine*, Elsevier Publishing Co. (Amsterdam), 1972.

T. C. Ruch and H. D. Patton, *Physiology and Biophysics*, Vol. II (Circulation, Respiration, and Fluid Balance), 20th ed, W. B. Saunders (Philadelphia), 1974.

W. Simon, *Mathematical Techniques for Physiology and Medicine*, Academic Press (N.Y.), 1972.

CHAPTER 9

Parameter Estimation in Two-Compartment Models

9.1 INTRODUCTION

A general principle in constructing a compartment model for a physical system is to select the model with the fewest number of compartments that adequately describes the data available on the system. It is quite remarkable that in physiological and pharmaceutical applications, models with two or three comparments often do an excellent job in describing the kinetics. Shown in Fig. 9.1 is a two-compartment model that occurs frequently in drug kinetics.

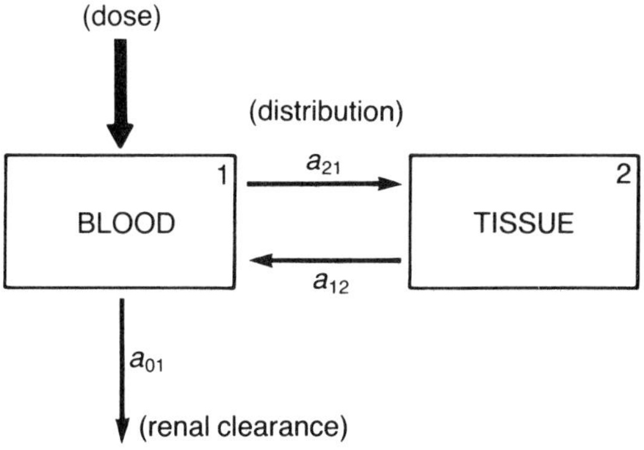

Fig. 9.1: A common two-compartment model for tracer kinetics

In many cases, it is possible to obtain tracer concentration measurements from one compartment only (typically from the blood or urine). We wish therefore to investigate under what conditions these measurements are sufficient to obtain estimates for the turnover rates a_{ij}, the volumes V_j, and the flow rates F_{ij}. As we will see, measurements from one compartment are rarely adequate when the number of compartments exceeds two. Although a four-or five-compartment model may appear to be reasonable in view of the physiology, the model may be totally useless from a clinical point of view unless the technology is present for sampling more compartments.

Shown in Fig. 9.2 is the bath-tub representation of the general two-compartment model. We will suppose that tracer makes its way into the system through the first compartment only so that $\mathbf{X}(0) = [x_0, 0]$ and $\mathbf{I}(t) = [I_1(t), 0]$.

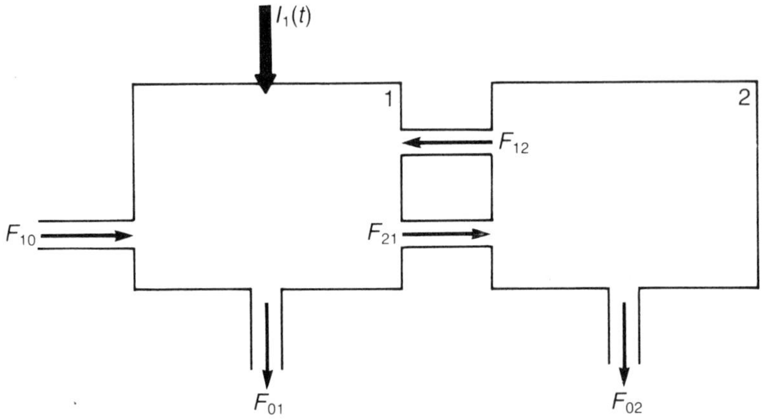

Fig. 9.2: Infusion into the first compartment

Hence

$$\dot{x}_1 = -(F_{21} + F_{01})c_1(t) + F_{12}c_2(t) + I_1(t)$$

$$\dot{x}_2 = F_{21}c_1(t) - (F_{12} + F_{02})c_2(t)$$

or, in terms of $\mathbf{X} = [x_1, x_2]$ alone,

$$\dot{\mathbf{X}} = \begin{bmatrix} -(a_{21} + a_{01}) & a_{12} \\ a_{21} & -(a_{12} + a_{02}) \end{bmatrix} \mathbf{X} + \begin{bmatrix} I_1(t) \\ 0 \end{bmatrix}$$

where $a_{ij} = F_{ij}/V_j$, and the original diagram takes the simple form shown in Fig. 9.3.

9.2 THE HOMOGENEOUS CASE

We will first analyze the case where tracer is rapidly injected into the first compartment. Hence $\mathbf{X}(0) = [x_0, 0]$ but $I_1(t) \equiv 0$.

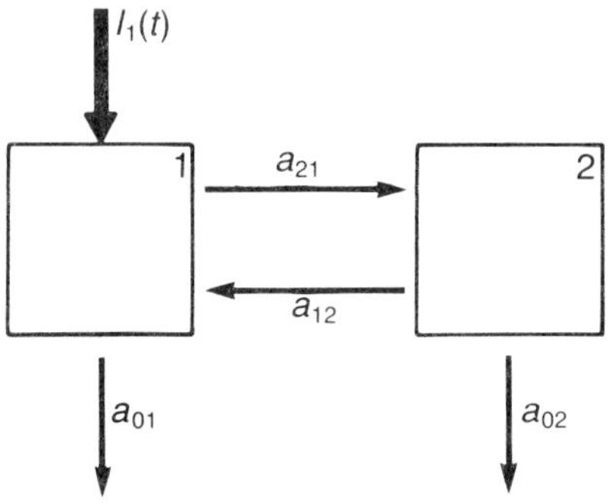

Fig. 9.3: Simplified two-compartment model with transfer coefficients

Let $a = a_{21} + a_{01}$, $b = a_{12}$, $c = a_{21}$, and $d = a_{12} + a_{02}$. The homogeneous solution $\mathbf{X}_c(t)$ is given in terms of the eigenvalues and eigenvectors of the matrix

$$\mathbf{A} = \begin{bmatrix} -a & b \\ c & -d \end{bmatrix}$$

The characteristic polynomial of **A** is $\lambda^2 + (a + d)\lambda + (ad - bc)$ with discriminant

$$D = (a + d)^2 - 4(ad - bc) = (a - d)^2 + 4bc.$$

Hence $D \geqslant 0$ and so both eigenvalues are real. Also note that $ad - bc = a_{21}a_{02} + a_{01}a_{12} + a_{01}a_{02} \geqslant 0$ and $a + d = a_{12} + a_{02} + a_{21} + a_{01} > 0$. Thus, using the quadratic formula, we have established the following theorem.

Theorem 9.1 In the general two-compartment model, both eigenvalues of the transfer matrix **A** are real and non-positive.

Let $\lambda_1 = -m_1$ and $\lambda_2 = -m_2$ be the two eigenvalues, where m_1 and $m_2 \geqslant 0$. There is the possibility that $\lambda_1 = \lambda_2$ but this will occur only when $a = d$ and $bc = 0$. If, for example, $a_{12} = b = 0$, we would then need $a_{02} = a_{21} + a_{01}$. Since this case is highly unlikely in applications, we will assume that *the two eigenvalues are distinct*. The solutions to $\dot{\mathbf{X}} = \mathbf{A}\,\mathbf{X}$ are now of the form

$$\mathbf{X}(t) = \mathbf{E}_1\, e^{-m_1 t} + \mathbf{E}_2\, e^{-m_2 t}$$

where $\mathbf{E}_1 = [c_1, c_2]$ and $\mathbf{E}_2 = [d_1, d_2]$ are eigenvectors corresponding to λ_1 and λ_2 respectively. From $\mathbf{A}\,\mathbf{E}_1 = \lambda_1 \mathbf{E}_1$, $\mathbf{A}\,\mathbf{E}_2 = \lambda_2 \mathbf{E}_2$, and $\mathbf{X}(0) = [x_0, 0]$, we obtain the equations

$$(a - m_1)c_1 = bc_2 \tag{1}$$

$$(d - m_1)c_2 = cc_1 \tag{2}$$

$$(a - m_2)d_1 = bd_2 \tag{3}$$

$$(d - m_2)d_2 = cd_1 \tag{4}$$

$$c_1 + d_1 = x_0 \tag{5}$$

$$c_2 + d_2 = 0 \tag{6}$$

In addition, since $\lambda^2 + (a + d)\lambda + (ad - bc) = (\lambda + m_1)(\lambda + m_2)$, we have

$$m_1 m_2 = ad - bc \tag{7}$$

$$m_1 + m_2 = a + d \tag{8}$$

Adding (2) and (4) we obtain

$$(d - m_1)c_2 + (d - m_2)d_2 = cx_0$$

Since $c_2 + d_2 = 0$, it follows that $c_2 = cx_0/(m_2 - m_1)$ and $d_2 = -c_2$. From (2), $c_1 = (d - m_1)c_2/c = x_0(d - m_1)/(m_2 - m_1)$. Finally, from (4), $d_1 = (d - m_2)d_2/c = x_0(m_2 - d)/(m_2 - m_1)$. We have established the following result.

Theorem 9.2 If $\lambda_1 \neq \lambda_2$, the solutions to $\dot{\mathbf{X}} = \mathbf{A}\,\mathbf{X}$ with $\mathbf{X}(0) = [x_0, 0]$ are

$$x_1(t) = \frac{x_0}{m_2 - m_1}\left[(d - m_1)\,e^{-m_1 t} + (m_2 - d)\,e^{-m_2 t}\right]$$

$$x_2(t) = \frac{x_0 c}{m_2 - m_1}\left[e^{-m_1 t} - e^{-m_2 t}\right]$$

where $m_1 = -\lambda_1$, $m_2 = -\lambda_2$, $d = a_{12} + a_{02}$, and $c = a_{21}$.

Corollary 9.2.1 The maximum value of $x_2(t)$ occurs at time $T = (m_2 - m_1)^{-1} \ln(m_2/m_1)$.

Returning to the special compartment model in Fig. 9.1, let $x_3(t)$ = amount of drug eliminated at time t. Then $\dot{x}_3(t) = a_{01}x_1(t)$. Assuming $x_3(0) = 0$, we have, using formula (7),

$$x_3(t) = \frac{a_{01}x_0}{m_2 - m_1}\left[\frac{m_1 - d}{m_1}\,e^{-m_1 t} + \frac{d - m_2}{m_2}\,e^{-m_2 t}\right] + x_0$$

and $\lim_{t\to\infty} x_3(t) = x_0$, as expected.

Next assume that we have obtained sample concentrations

$$\overset{*}{c}_1(t_1), \overset{*}{c}_1(t_2), \ldots, \overset{*}{c}_1(t_n)$$

from the first compartment. Call all of the transfer coefficients be estimated from this data alone? Shown in Fig. 9.4 is the graph of typical data.

We must find a function of the form $A\,e^{-\alpha t} + B\,e^{-\beta t}$ that fits the data. If $\beta > \alpha$, then, for large t, $c_1(t) \approx A\,e^{-\alpha t}$. Hence $\ln c_1(t) \approx \ln A - \alpha t$. By fitting a line to the tail end of the data $\ln \overset{*}{c}_1(t)$, we can obtain estimates for A and α.

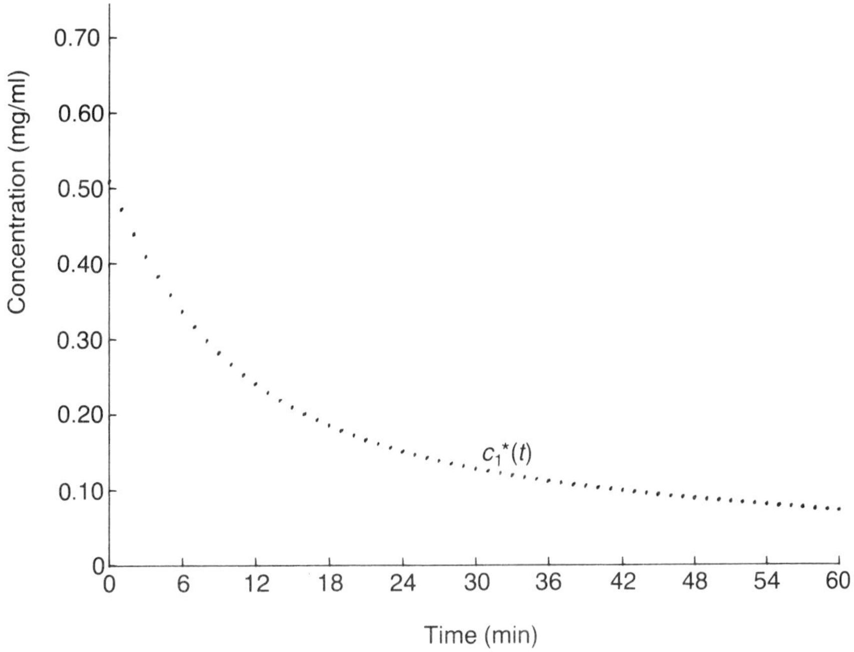

Fig. 9.4: Drug concentration data

Shown in Fig. 9.5 is the original data plotted on 'log paper'. We obtain $A = 0.2$ and $\alpha = 0.0025$.

We then plot ln $[c_1^*(t) - A\ \mathrm{e}^{-\alpha t}]$ versus t, which should be approximately linear with slope $-\beta$ and intercept ln B. Again, from Fig. 9.5, we obtain $B = 0.32$ and $\beta = 0.01$. Hence

$$c_1^*(t) \approx 0.2\ \mathrm{e}^{-0.0025t} + 0.32\ \mathrm{e}^{-0.01t}$$

This curve fitting method is called the *method of exponential peeling*. (Another commonly used method is based on the method of least squares.)

From the solution given in Theorem 9.2, we obtain estimates for m_1, m_2, $c_0(d - m_1)/(m_2 - m_1)$, and $c_0(m_2 - d)/(m_2 - m_1)$. Since $c_0 = c_1(0)$ is known, we can obtain an estimate for $d = a_{12} + a_{02}$.

From equation (8) on p. 130, $a = m_1 + m_2 - d$ and so an estimate for $a = a_{21} + a_{01}$ results. From equation (7), $bc = a_{12}a_{21} = ad - m_1 m_2$. Hence we may estimate $a_{12}a_{21}$ but in general it is not possible to solve for either a_{12} or a_{21} without further information.

To illustrate this point, we will use the data from Fig. 9.4. Now $c_0 = c_1(0) \approx 0.52$ and we obtain

$$d = a_{12} + a_{02} \approx 0.0054$$

$$a = a_{21} + a_{01} \approx 0.0071$$

$$a_{12}a_{21} \approx 0.0000133$$

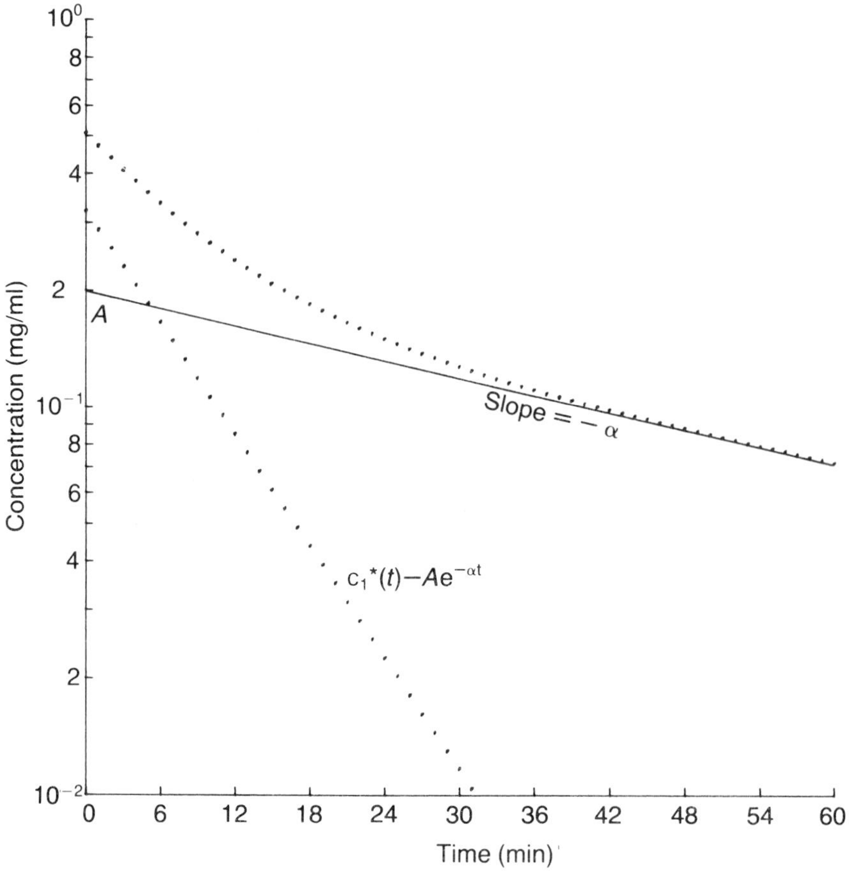

Fig. 9.5: Method of exponential peeling

This system of *three equations in four unknowns* does not have a unique solution. There are, however, two commonly occurring special cases for which sampling from one compartment is sufficient.

Case I: $a_{02} = 0$

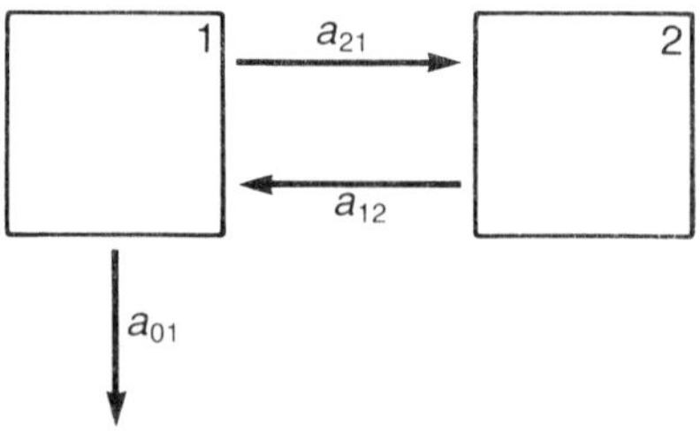

In this case $d = a_{12}$ will be known and so a_{21} can be estimated from $a_{21} = bc/a_{12}$ and then $a_{01} = a - a_{21}$. For the given data, $a_{12} \approx 0.0054$, $a_{21} \approx 0.0025$, and $a_{01} \approx 0.0046$.

Case I: $a_{01} = 0$

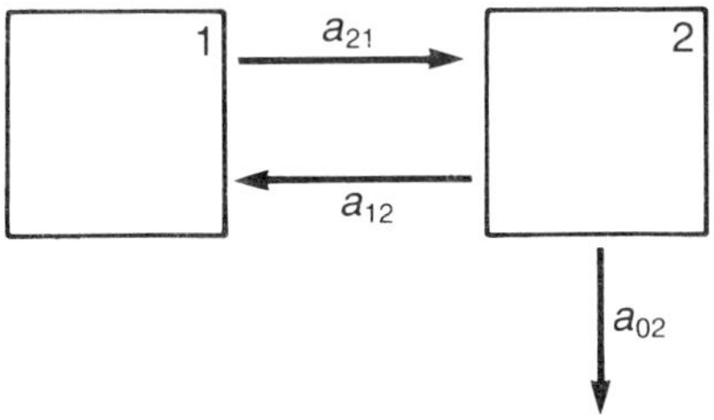

For this special case, $a = a_{21}$ will be known, a_{12} can be estimated from $a_{12} = bc/a_{21}$ and then $a_{02} = d - a_{12}$ can be estimated. For the given data, $a_{21} \approx 0.0071$, $a_{12} \approx 0.0019$, and $a_{02} \approx 0.0035$.

Note that when $a_{12} = 0$, $d = a_{02}$ will be known as will be $a = a_{21} + a_{01}$, but there is insufficient information to find a_{21} or a_{01}. A similar situation holds if $a_{21} = 0$.

Example 9.1 Creatinine is a component of urine originally formed from creatinine phosphate, a substance which supplies energy to the muscles. In an experiment carried out by Sapirstein *et al.* [6], $x_0 = 2$ grams of creatinine were injected into the bloodstream of dogs and subsequent concentrations in the bloodstream were found to be fitted by

$$c_1(t) = 0.188\, e^{-0.0161t} + 0.321\, e^{-0.1105t}.$$

Assuming the model in Fig. 9.1 to be appropriate, estimate a_{12}, a_{21}, a_{01}, V_1, and V_2.

Solution 9.1 We are given that $m_1 \approx 0.0161$, $m_2 \approx 0.1105$, and $c_1(0) = 0.509 = x_0/V_1$. Hence $V_1 = 2/0.509 = 3.93$ liters. Since $x_1(t) = c_1(t)\, V_1$, it follows that

$$x_1(t) = 0.7387\, e^{-0.0161t} + 1.2613\, e^{-0.1105t}$$

Hence, $x_0(d - m_1)/(m_2 - m_1) = 0.7387$ or $d = a_{12} = 0.051$. It follows that $a = m_1 + m_2 - d = 0.0756$ and $a_{12}a_{21} = ad - m_1 m_2 = 0.0021$. Therefore $a_{21} = 0.0407$ and $a_{01} = a - a_{21} = 0.0349$ per minute.

Finally, to estimate V_2, note that $F_{12} = F_{21}$. Hence $V_2 = F_{12}/a_{12} = F_{21}/a_{12} = (a_{21}/a_{12})\, V_1 \approx 3.136$ liters.

9.3 THE NON-HOMOGENEOUS CASE

In clinical applications it is often desirable to maintain a constant therapeutic level of a particular drug in a patient. To accomplish this, a drug is administered at a constant rate of I mg/minute through intravenous infusion into the bloodstream. The techniques of Chapter 7 can then be used to solve

$$\dot{\mathbf{X}} = \mathbf{A}\,\mathbf{X} + \begin{bmatrix} I \\ 0 \end{bmatrix}$$

and find the steady state solution $\hat{\mathbf{X}}$. Assuming that both eigenvalues of $\mathbf{A}$ are negative (which will occur when a_{01} and $a_{12} > 0$), the constant particular solution $\mathbf{X}_p$ is given by

$$\mathbf{X}_p = -\mathbf{A}^{-1} \begin{bmatrix} I \\ 0 \end{bmatrix}$$

$$= \frac{1}{ad - bc} \begin{bmatrix} d & b \\ c & a \end{bmatrix} \begin{bmatrix} I \\ 0 \end{bmatrix} = \frac{1}{ad - bc} \begin{bmatrix} Id \\ Ic \end{bmatrix}$$

Since $\lim_{t \to \infty} \mathbf{X}_c(t) = 0$, the steady state solution $\hat{\mathbf{X}}$ is just $\mathbf{X}_p$.

Returning to the special case illustrated in Fig. 9.1, we have $a_{02} = 0$ and so $ad - bc = a_{12}a_{01}$, $d = a_{12}$, and $c = a_{21}$. The steady state solution is now given by

$$\hat{x}_1 = I/a_{01} \quad \text{and} \quad \hat{x}_1 = (a_{21}/a_{12})\,\hat{x}_1$$

The *steady state concentration in the bloodstream* $\hat{c}_1(t)$ is therefore given by

$$\boxed{\hat{c}_1 = I/(a_{01} V_1)}$$

Example 9.2 The drug *lidocaine* is frequently used to treat ventricular arrhythmias (irregular heartbeat) and is also widely used as a local anesthetic. It is well known that the model illustrated in Fig. 9.1 does an effective job of simulating lidocaine kinetics (see [4]). Normal turnover rates are $a_{01} = 0.024$ per minute, $a_{12} = 0.038$ per minute, and $a_{21} = 0.066$ per minute, and the volume of distribution is about 30 liters. In the treatment of arrhythmias, the minimum therapeutic level is a concentration of 1.5 mg/liter in the bloodstream, and concentration levels above 6 mg/liter produce serious side-effects which can result in death. We wish therefore to find a drug–dose scheme which reaches that minimum therapeutic level quickly but avoids toxic concentration levels.

In order to achieve a steady state concentration of 3.5 mg/liter in the bloodstream, the infusion rate must be $I = (a_{01}V_1)\hat{c}_1 = (0.024)\,(30)\,(3.5) = 2.52$ mg/min.

(a) Under these conditions, find how long it takes to reach the minimum therapeutic level of 1.5 mg/liter if there is *no load dose*.

(b) If in addition to the infusion rate of 2.52 mg/min there is a load dose of 100 mg, find the minimum and maximum concentration in the plasma, and find the maximum amount of lidocaine in the tissues.

Solution 9.2 The amount of lidocaine in the two compartments satisfies the differential equation $\dot{\mathbf{X}} = \mathbf{A}\,\mathbf{X} + [I, 0]$ where

$$\mathbf{A} = \begin{bmatrix} -0.090 & 0.038 \\ 0.066 & -0.038 \end{bmatrix}$$

The steady state solution is $\mathbf{X}_p = [I/a_{01}, (a_{21}/a_{12})\,I/a_{01}] = I\,[41.667, 72.368]$, and the complementary solution is

$$\mathbf{X}_c(t) = c_1 \begin{bmatrix} 0.73039 \\ -0.58483 \end{bmatrix} \mathrm{e}^{-0.12043t} + c_2 \begin{bmatrix} 0.26961 \\ 0.58483 \end{bmatrix} \mathrm{e}^{-0.00757t}$$

(a) If $\mathbf{X}(0) = \mathbf{0}$ ('no load dose') then we must find constants c_1 and c_2 so that $\mathbf{X}_c(0) = -\,\mathbf{X}_p$. The required solution is $c_1 = -8.3048\,I$ and $c_2 = -132.047\,I$. For the infusion rate of $I = 2.52$ mg/min, it follows that

$$x_1(t) = -15.285\,\mathrm{e}^{-0.12043t} - 89.715\,\mathrm{e}^{-0.00757t} + 105$$

To find when $c_1(t) = 1.5$ mg/liter, we solve $x_1(t) = 45$. Using Newton's method, for example, we find that $t \approx 53.2$ minutes.

(b) To find the solution $\mathbf{X}(t)$ satisfying $\mathbf{X}(0) = [100, 0]$, all we need do is take the solution $\mathbf{Y}(t)$ in (a) satisfying $\mathbf{Y}(0) = \mathbf{0}$ and add the solution

$$\mathbf{Z}(t) = 100 \begin{bmatrix} 0.73039 \\ -0.58483 \end{bmatrix} \mathrm{e}^{-0.12043t} + 100 \begin{bmatrix} 0.26961 \\ 0.58483 \end{bmatrix} \mathrm{e}^{-0.00757t}$$

from Theorem 9.2 which satisfies $\mathbf{Z}(0) = [100, 0]$. Hence, letting $\mathbf{X}(t) = \mathbf{Y}(t) + \mathbf{Z}(t)$, we have

$$x_1(t) = 57.754\,\mathrm{e}^{-0.12043t} - 62.754\,\mathrm{e}^{-0.00757t} + 105$$

$$x_2(t) = -46.244\,\mathrm{e}^{-0.12043t} - 136.124\,\mathrm{e}^{-0.00757t} + 182.368$$

Setting $\dot{x}_1(t) = 0$, we obtain $t = 23.8$ minutes. The minimum value of $x_1(t)$ is then 55.88 mg, and so the minimum concentration in the bloodstream is 55.88 mg/30 liters = 1.86 mg/liter. The graph of $c_1(t)$ is shown in Fig. 9.6. Note that *at all times the concentration lies between 1.86 mg/liter and 3.5 mg/liter.*

Since $\dot{x}_2(t) > 0$, the amount of lidocaine in the tissues increases from 0 mg to the steady state level of 182.368 mg.

9.4 PARAMETER ESTIMATION IN MULTI-COMPARTMENTAL MODELS

The general solution to $\dot{\mathbf{X}} = \mathbf{A}\,\mathbf{X}$ with $\mathbf{X}(0) = \mathbf{X}_0$ usually takes the form

$$\mathbf{X}(t) = \mathbf{E}_1\,\mathrm{e}^{-m_1 t} + \mathbf{E}_2\,\mathrm{e}^{-m_2 t} + \ldots + \mathbf{E}_n\,\mathrm{e}^{-m_n t}$$

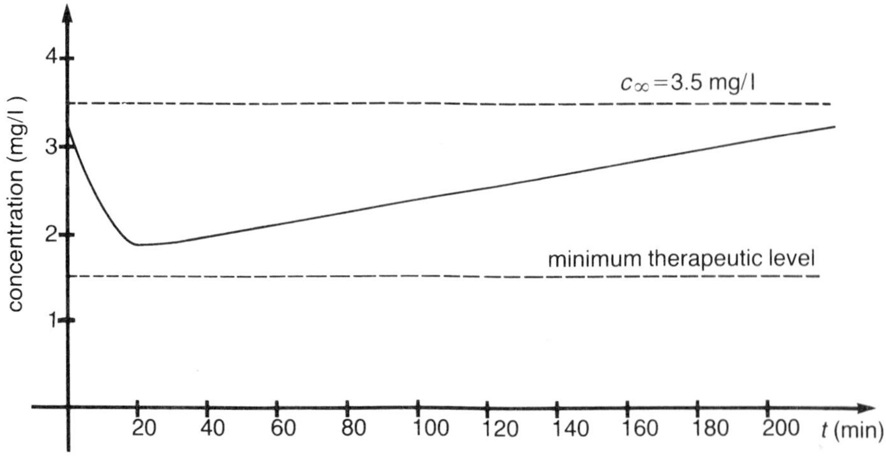

Fig. 9.6: Lidocaine concentration curve

where $m_k \geqslant 0$, for $k = 1$ to n, and $\mathbf{E}_k = \mathbf{E}_k(\mathbf{A}, \mathbf{X}_0)$ is an eigenvalue corresponding to $\lambda_k = -m_k$ whose form depends on the unknown transfer coefficients $a_{11}, a_{12}, \ldots, a_{nn}$ and $\mathbf{X}_0$.

If eigenvectors $\mathbf{E}_1, \ldots, \mathbf{E}_n$ and eigenvalues $\lambda_1, \ldots, \lambda_n$ are known, then the transfer matrix $\mathbf{A}$ can be recaptured using the spectral decomposition theorem. We let $\mathbf{Z} = [\mathbf{E}_1 \,\vdots\, \ldots \,\vdots\, \mathbf{E}_n]$, construct the spectral components $\mathbf{Z}_k$ (see the proof of Theorem 3.1), and then

$$\mathbf{A} = \lambda_1 \mathbf{Z}_1 + \lambda_2 \mathbf{Z}_2 + \ldots + \lambda_n \mathbf{Z}_n$$

If we sample *from one compartment only* (say compartment one) and fit a function of the form

$$b_1\, e^{-\alpha_1 t} + b_2\, e^{-\alpha_2 t} + \ldots + b_n\, e^{-\alpha_n t}$$

then we obtain estimates for the eigenvalues of $\mathbf{A}$ but have approximations for *only the first component* of each eigenvector $\mathbf{E}_k$.

If we denote the first component of $\mathbf{E}_k$ by $z_{1k} = z_{1k}(a_{11}, a_{12}, \ldots, a_{nn})$ then we have the system of n equations

$$\begin{aligned} z_{11}(a_{11}, a_{12}, \ldots, a_{nn}) &= b_1 \\ z_{12}(a_{11}, a_{12}, \ldots, a_{nn}) &= b_2 \\ &\vdots \\ z_{1n}(a_{11}, a_{12}, \ldots, a_{nn}) &= b_n \end{aligned}$$

Since $\mathbf{X}_0$ will be known, the last equation is actually redundant since $\mathbf{E}_n = \mathbf{X}_0 - (\mathbf{E}_1 + \mathbf{E}_2 + \ldots + \mathbf{E}_{n-1})$. In addition, the characteristic polynomial is now known to be $(\lambda + m_1)(\lambda + m_2) \ldots (\lambda + m_n)$. Since

$$p(\lambda) = \lambda^n - (a_{11} + a_{22} + \ldots + a_{nn})\, \lambda^{n-1} + \ldots + (-1)^n \det \mathbf{A}$$

we obtain an additional n equations in n^2 unknowns by equating the coefficients in the two expressions for $p(\lambda)$.

The problem therefore reduces to solving $(n - 1) + n = 2n - 1$ equations in the n^2 unknown $a_{11}, a_{12}, \ldots, a_{nn}$, and, in general, *solutions will not be unique.* As the next example illustrates, sampling from one compartment is not always sufficient even when most of the entries in $\mathbf{A}$ are known to be zero.

Example 9.3 Fig. 9.7 is a simple three-compartment model in which there are only three unknown transfer coefficients $a_{21} = \lambda_1$, $a_{32} = \lambda_2$, and $a_{03} = \lambda_3$. If tracer is quickly injected into the first compartment then $\mathbf{X}(0) = [x_0, 0, 0]$ and it is not hard to show that

$$x_1(t) = x_0\, e^{-\lambda_1 t}$$

$$x_2(t) = \frac{\lambda_1 x_0}{\lambda_2 - \lambda_1}\,[e^{-\lambda_1 t} - e^{-\lambda_2 t}]$$

$$x_3(t) = \frac{\lambda_1 \lambda_2 x_0}{\lambda_2 - \lambda_1}\left[\frac{1}{\lambda_3 - \lambda_1}\, e^{-\lambda_1 t} - \frac{1}{\lambda_3 - \lambda_2}\, e^{-\lambda_2 t} + \frac{\lambda_2 - \lambda_1}{(\lambda_3 - \lambda_1)(\lambda_3 - \lambda_2)}\, e^{-\lambda_3 t}\right]$$

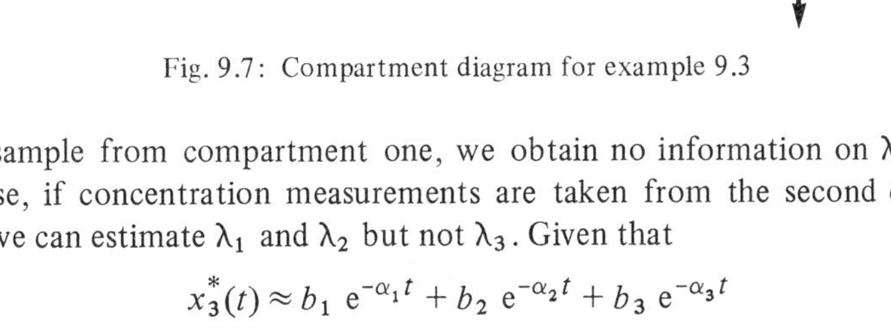

Fig. 9.7: Compartment diagram for example 9.3

If we sample from compartment one, we obtain no information on λ_2 or λ_3. Likewise, if concentration measurements are taken from the second compartment, we can estimate λ_1 and λ_2 but not λ_3. Given that

$$\overset{*}{x}_3(t) \approx b_1\, e^{-\alpha_1 t} + b_2\, e^{-\alpha_2 t} + b_3\, e^{-\alpha_3 t}$$

we obtain estimates for λ_1, λ_2, and λ_3 but we do not know, e.g., whether λ_1 is α_1, α_2, or α_3. Since x_0 will be known, we can test all six possibilities for λ_1, λ_2, and λ_3 and compare b_1, b_2, and b_3 with the coefficients in the expression for $x_3(t)$.

Fig. 9.8 shows a three-compartment model which occurs frequently in pharmacological applications. It can be shown that all transfer coefficients can be estimated given data $x_1^*(t)$ from the central comparment. Such a system, in which there are two-way flows between a central compartment and any other compartment, is called a *mammilary system*. For details, see [2], pages 92–103.

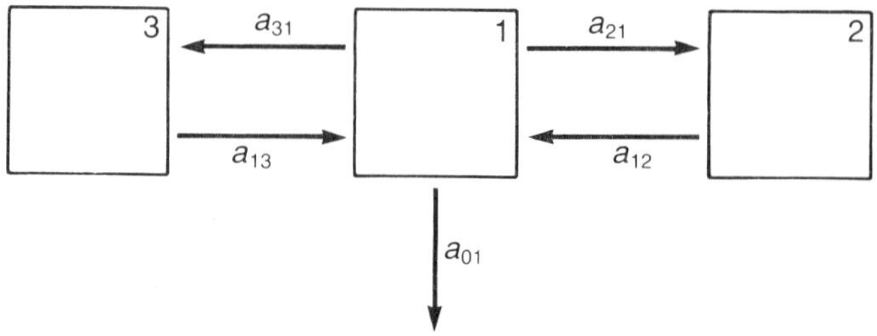

Fig. 9.8: A three-compartment mammilary system

EXERCISES

Part A

1 (a) For the general two-compartment model, find the solution to $\dot{\mathbf{X}} = \mathbf{A}\,\mathbf{X}$ satisfying $\mathbf{X}(0) = [0, y_0]$.
(b) Use the results of (a) and Theorem 9.2 to find the solution to $\dot{\mathbf{X}} = \mathbf{A}\,\mathbf{X}$ satisfying $\mathbf{X}(0) = [x_0, y_0]$.

2 (a) Find the solution to the two-compartment model shown below that satisfies $\mathbf{X}(0) = \mathbf{0}$.

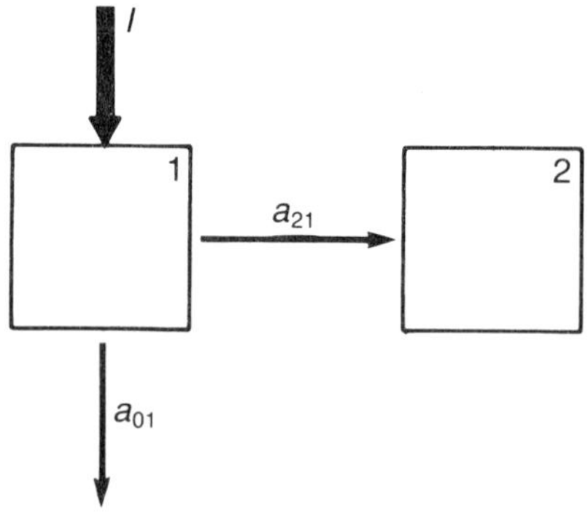

(b) Discuss how data from compartment two can be fitted by a solution curve.
(c) In a model for the incorporation of labeled thymine into the DNA of *E. coli* during replication, the two compartments are (1) labeled thymine precursor and (2) labeled thymine in DNA [5]. We assume that the influx of labeled thymine into the precursor pool is constant. From precursor experiments due to R. Werner [8],

$$x_2(t) \approx 36\,600\,(0.27t - 1 + e^{-0.27t})$$

Using this information, estimate $x_1(t)$.

3 To study the movement of potassium in the bloodstream between the plasma and red blood cells, a fixed quantity x_0 of radioactive ^{42}K is injected into the bloodstream. Assuming that ^{42}K is not lost to the system, then the compartment model shown below is appropriate.

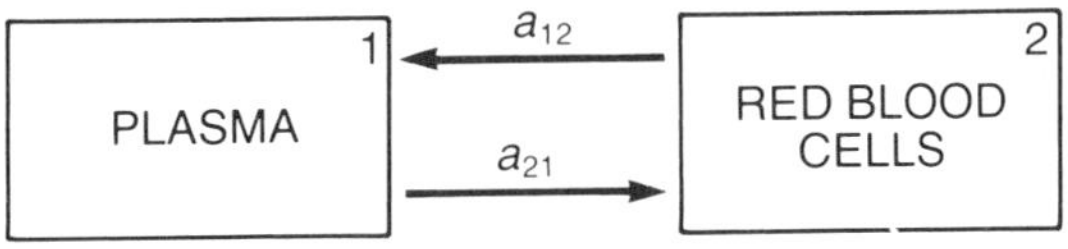

(a) Solve the resulting differential equation given that $\mathbf{X}(0) = [x_0, 0]$.

(b) Find $\hat{\mathbf{X}} = \lim_{t\to\infty} \mathbf{X}(t)$.

(c) Shown in the table below is plasma data taken from an experiment of Sheppard and Martin [7]. Use the method of exponential peeling to estimate $x_1(t)/x_0$.

t (min)	0	60	124	181	300	420	540	1270
Relative specific activity $x_1(t)/x_0$	1.0	0.73	0.52	0.41	0.25	0.175	0.123	0.065

(d) Estimate a_{12} and a_{21}.

4 Shown in the figure below is a compartmental representation of the cell membrane, the extracellular fluid, and the fluid within the cell.

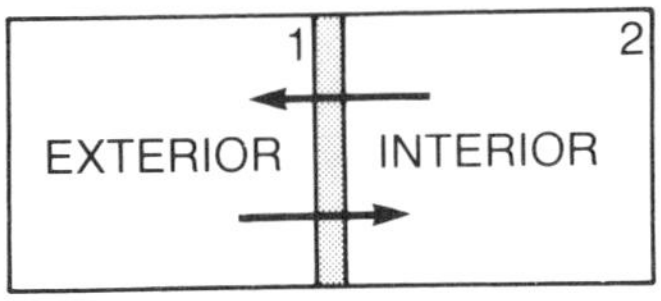

Materials necessary for cell growth and maintenance must make their way past the membrane. One of the transport mechanisms, known as *passive diffusion*, is governed by Fick's law: the rate of transport is directly proportional to the concentration gradient.

(a) Show that $\dot{x}_1 = \kappa(c_2 - c_1)$ and $\dot{x}_2 = \kappa(c_1 - c_2)$ for some constant $\kappa > 0$ (called the *permeability constant*).

(b) Solve for $c_1(t)$ and $c_2(t)$ given that $c_1(0) = a$ and $c_2(0) = b$.

(c) Find $\hat{\mathbf{C}} = \lim_{t\to\infty} \mathbf{C}(t)$.

(Adapted from M. R. Cullen, *Mathematics for the Biosciences*, PWS Publishers, 1983.)

5 A 2.5 mg load dose of a drug is administered to a patient. Resulting plasma concentrations of the drug are fitted by the curve

$$c_1(t) = 27\, e^{-0.32t} + 4.4\, e^{-0.03t}\ \mu g/ml$$

where t is in minutes. Assuming that the model in Fig. 9.1 is appropriate, estimate a_{12}, a_{21}, a_{01}, V_1, and V_2.

6 Fit the sum of two exponential functions to the drug concentration data

t (hours)	1	2	3	4	5	7	9	12	15	18
$c(t)$ (mg/liter)	80	51	37	29	25	21	18	15	13	11

If $x_0 = 2$ grams, estimate all parameters assuming that $a_{02} = 0$.

7 (a) For the two-compartment model shown below, show that for the

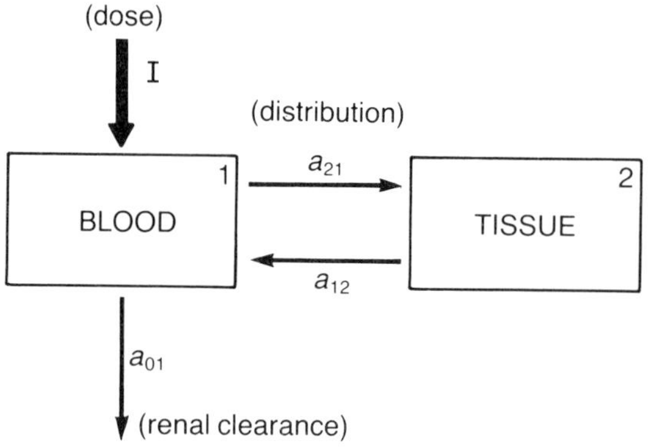

solution satisfying $\mathbf{X}(0) = [x_0, 0]$,

$$x_1(t) = \frac{I}{a_{01}} - \frac{(x_0 m_1 - I)(m_1 - a_{12})}{m_1(m_2 - m_1)} e^{-m_1 t} + \frac{(x_0 m_2 - I)(m_2 - a_{12})}{m_2(m_2 - m_1)} e^{-m_2 t}$$

(b) Show that when the relationship between the load dose x_0 and infusion rate I is $x_0 = I/m_2$, then $x_1(t)$ is either strictly increasing or strictly decreasing.

8 In the clinical application of the model in Exercise 7, we wish to combine a load dose x_0 and infusion rate I that results in constant drug concentration in the plasma being reached as quickly as possible and that avoids toxic concentration levels. The following scheme has been suggested [3]. Let m_2 be the smaller of the two eigenvalues. Then if $\hat{c}_1$ is the desired concentration level,

set $I = a_{01}\ V_1\ \hat{c}_1$ and $x_0 = I/m_2$. Carry out this scheme for the drug *lidocaine* using the parameters in Example 9.2 and $\hat{c}_1 = 3.5$ mg/liter.

(a) What is the infusion rate and load dose?

(b) Is the toxic concentration of 6 mg/liter ever exceeded? Sketch the graph of $c_1(t)$.

(c) How long does it take to get within 0.1 mg/liter of the equilibrium concentration?

9 The drug *theophylline* is used in the treatment of asthma. When maintained in the bloodstream at levels between 10–20 μg/ml, the drug gives impressive results in alleviating the symptoms of chronic asthma. When concentrations exceed 20 μg/ml, the toxic effects can be severe. A preliminary experiment was performed on a patient with a load dose of 140 mg. Plasma concentrations were fitted by

$$c_1(t) = 13.13\ e^{-4.89t} + 5.38\ e^{-0.183t}\ \mu g/ml$$

where t is in hours.

(a) Estimate a_{12}; a_{21}, a_{01}, and V_1.

(b) Apply the drug–dose scheme in Exercise 8 to find the load dose and infusion rate for a steady state concentration of 15 μg/ml.

(c) Sketch the graph of $c_1(t)$. Are toxic levels ever reached?

(d) How long does it take to get within 1 μg/ml of the steady state concentration?

For each of the following two-compartment models, find the solution which satisfies the initial condition $\mathbf{X}(0) = [x_0, 0]$.

10

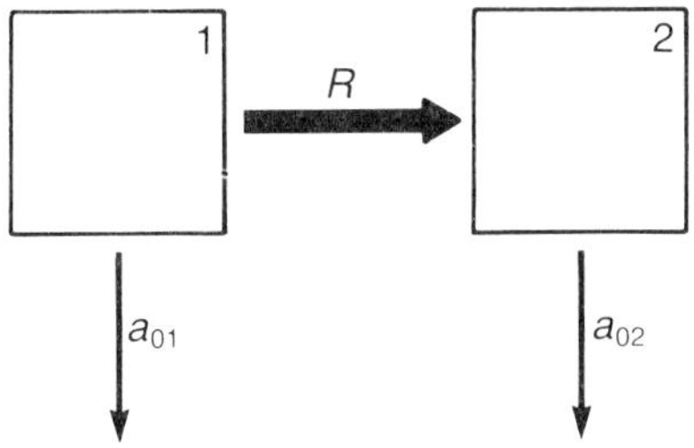

11

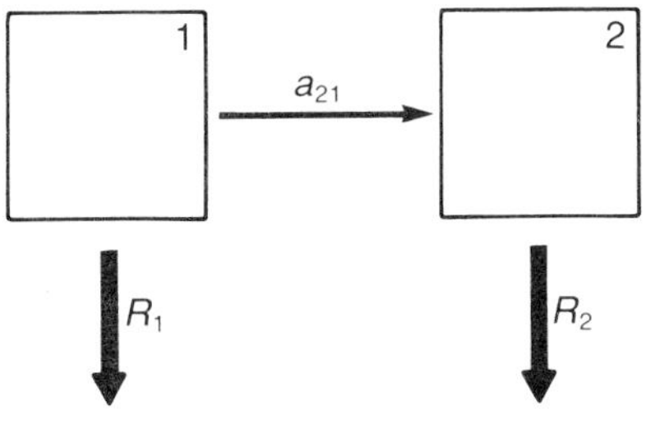

12

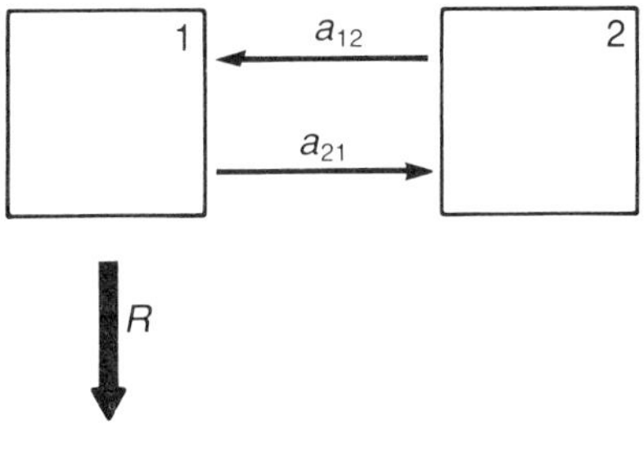

Part B

13 The following three-compartment model has been proposed when a drug is taken orally.

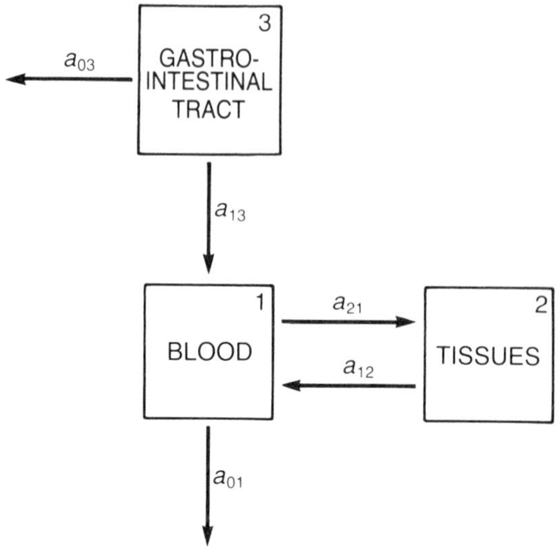

(a) Show that $x_3(t) = x_0\ \mathrm{e}^{-(a_{03}+a_{13})t}$ and if $I_1(t) = a_{13}\ x_0\ \mathrm{e}^{-(a_{03}+a_{13})t}$, then

$$\dot{\mathbf{X}} = \mathbf{A}\ \mathbf{X} + [I_1(t), 0]$$

where $\mathbf{X} = [x_1, x_2]$.

(b) Find a particular solution $\mathbf{X}_p(t) = \mathbf{C}\ \mathrm{e}^{-(a_{03}+a_{13})t}$, using Theorem 7.3. Then find the solution satisfying $\mathbf{X}(0) = \mathbf{0}$.

(c) Following a 500 mg dose of lidocaine, plasma concentration data was fitted to the sum of three exponentials using the method of exponential peeling. Determine as many parameters as possible if

$$c_1(t) = -12\ \mathrm{e}^{-0.15t} + 20.2\ \mathrm{e}^{-0.015t} - 8.2\ \mathrm{e}^{-0.075t}$$

14 When a dose of x_0 mg is given, the concentration of drug in the bloodstream is estimated to be

$$c_1(t) = x_0\ [0.08\ \mathrm{e}^{-0.5t} + 0.055\ \mathrm{e}^{-0.04t}]\ \mathrm{mg/l}$$

(a) If x_0 mg are given very τ hours, find the expression for the maximum drug concentration in the bloodstream. (*Hint*: Let $a = x(\tau)/x_0$ and see Exercise 11, Chapter 8.)

(b) If individual doses are 100 mg and the maximum safe concentration is 35 mg/liter, how frequently can the drug be safely adminstered?

(c) If doses are to be given every 6 hours, find the size of the dose that results in a maximum concentration of 25 mg/liter. What will the long range minimum concentration be?

15 Some drugs (such as the penicillin oxacillin) are either too poorly soluble or too irritating to be administered in a single bolus. Therefore it is not possible to estimate turnover rates by a single load dose experiment. Instead, the drug is slowly infused into the bloodstream until equilibrium concentrations are reached. Then, infusion is stopped and drug plasma concentrations are taken. (See [1] for more information.)

(a) Solve $\dot{\mathbf{X}} = \mathbf{A}\,\mathbf{X}$ for the two-compartment model in Fig. 9.1 subject to the initial condition

$$\mathbf{X}(0) = [\, I/a_{01},\, a_{21}I/(a_{01}a_{12})] \,.$$

(*Hint*: Use Exercise 1 and simplify.)

(b) The sum of two exponential functions was fitted to plasma concentrations of oxacillin, obtained after infusion ceased, giving

$$c_1(t) = 2.942\ \mathrm{e}^{-0.127t} + 6.758\ \mathrm{e}^{-0.027t}\ \mu\mathrm{g/ml}$$

where t is in minutes. The infusion rate was $I = 250$ mg/hour. Estimate $a_{01}, a_{12}, a_{21}, V_1$, and V_2.

16 The following is a three-compartment model for the distribution of the drug *kanamycin*. A fixed dose x_0 is injected into the hip from which the drug

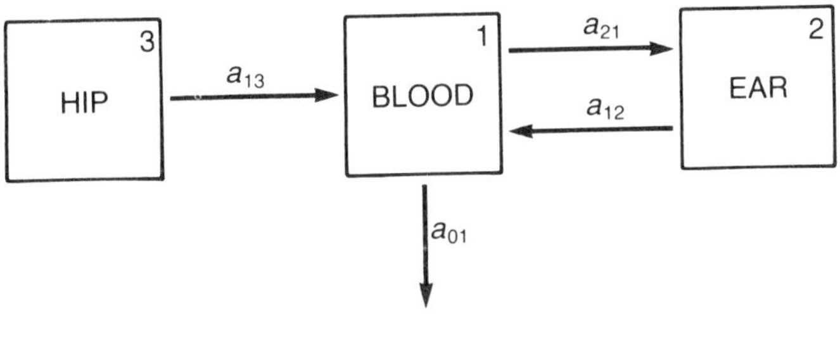

quickly diffuses into the bloodstream. We must monitor the concentration of kanamycin in the ear since too high a concentration can result in permanent hearing loss. We are interested in finding the maximum value of $x_2(t)$ in terms of x_0. The following approximate parameter values have been given for a 150 kg man: $V_1 = 18$ liters, $V_2 = 1$ ml, $V_3 = 2$ ml, $a_{13} = 2.2$/hour, $a_{12} = 0.06$/hour, and $a_{21} = 7.8 \times 10^{-7}$/hour.

(a) For a patient with normal kidney function, $a_{01} = 0.264$/hour. Find the maximum value of $c_2(t)$ in terms of x_0.

(b) For a patient with complete loss of kidney function, $a_{01} = 0.007$/hour. Find the maximum value of $c_2(t)$ in terms of x_0.

(c) If 500 mg doses are given every 8 hours to a normal patient, find $c_2(t)$ for $t = 8, 16, 24, \ldots, 8n, \ldots$ What is the maximum steady state concentration?

REFERENCES

[1] M. Gibaldi, 'Estimation of the Pharmacokinetic Parameters of the Two-Compartment Open Model from Post-Infusion Plasma Concentration Data', *J. Pharmaceutical Science* (1969), **58**(9), 1133–1135.

[2] M. Gibaldi and D. Perrier, *Pharmacokinetics* (2nd edn), Marcel Dekker, New York, 1982.

[3] P. A. Mitenko and R. I. Ogilve, 'Rapidly Achieved Plasma Concentration Plateaus, with Observations on Theophylline Kinetics', *Clinical Pharmacology and Therapeutics* (1972), **13**, 329–335.

[4] John H. Rodman, Chapter 11 (Lidocaine) in *Applied Pharmacokinetics*, William E. Evans, Jerome J. Schentag, and William J. Jusko (eds), Applied Therapeutics Inc., San Francisco, 1980.

[5] S. Rubinow and A. Yen. 'Quantitation of Some DNA Precursor Data', *Nature New Biology* (1972), **239**, 73–74.

[6] L. A. Sapirstein, D. G. Vidt, M. J. Mandel, and G. Hanusek, 'Volumes of Distribution and Clearances of Intravenously Injected Creatinine in the Dog', *American Journal of Physiology* (1955), **181**, 330–336.

[7] C. W. Sheppard and W. R. Martin, 'Cation Exchange between Cells and Plasma of Mammalian Blood. I. Methods and Application to Potassium Exchange in Human Blood', *J. General Physiology* (1950), **33**, 703–722.

[8] R. Werner, 'Nature of DNA Precursors', *Nature New Biology* (1971), **233**, 99–103.

CHAPTER 10

Approximations: Linearization and Numerical Methods

10.1 SYSTEMS OF DIFFERENTIAL EQUATIONS AND STATIONARY POINTS

For the n compartment model in which material flows continuously from one compartment to another, we let r_{ij} denote the *flux* from compartment j to compartment i. The hypothesis that $r_{ij} = a_{ij}\, x_j$ was called the *linear donor-controlled hypothesis* and lead to the system of differential equations $\dot{\mathbf{X}} = \mathbf{A}\,\mathbf{X}$, where $\mathbf{A}$ is the transfer matrix $[a_{ij}]$. When this simplifying assumption was made, we could apply our eigenvalue–eigenvector techniques to solve the system.

In general the flux rates r_{ij} will be functions of t and the states $x_1, \ldots, x_n$. We let $\mathbf{X} = [x_1, \ldots, x_n]$ and let $r_{ij} = r_{ij}(\mathbf{X}, t)$. It follows that

$$\dot{x}_i = \textit{net flux} \text{ into compartment } i$$

$$= \sum_{j \neq i} r_{ij}(\mathbf{X}, t) - \sum_{j \neq i} r_{ji}(\mathbf{X}, t)$$

Hence a given set of assumptions as to how material flows between compartments will lead to a system of differential equations of the form

$$\begin{aligned} \dot{x}_1 &= f_1(\mathbf{X}, t) \\ \dot{x}_2 &= f_2(\mathbf{X}, t) \\ &\vdots \\ \dot{x}_n &= f_n(\mathbf{X}, t) \end{aligned}$$

Letting $\mathbf{f}(\mathbf{X}, t) = [f_1(\mathbf{X}, t), f_2(\mathbf{X}, t), \ldots, f_n(\mathbf{X}, t)]$, the system can be written in the compact form $\dot{\mathbf{X}} = \mathbf{f}(\mathbf{X}, t)$. A *solution* $\mathbf{X} = \mathbf{X}(t)$ is a differentiable vector-valued function satisfying $\dot{\mathbf{X}}(t) = \mathbf{f}(\mathbf{X}(t), t)$.

The function $\mathbf{f}(\mathbf{X}, t)$ defines a *vector field* on n-dimensional Euclidean space $\mathbf{R}^n$. To be more specific, imagine that a wind is blowing over a region Ω in $\mathbf{R}^n$ and let $\mathbf{f}(\mathbf{X}, t)$ denote the velocity of the wind at time t and at a particular point $\mathbf{X}$ in Ω as illustrated in Fig. 10.1.

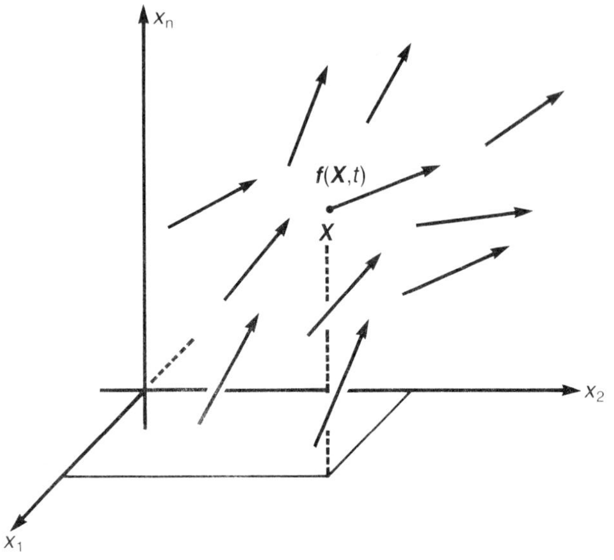

Fig. 10.1(a): The vector field at time t

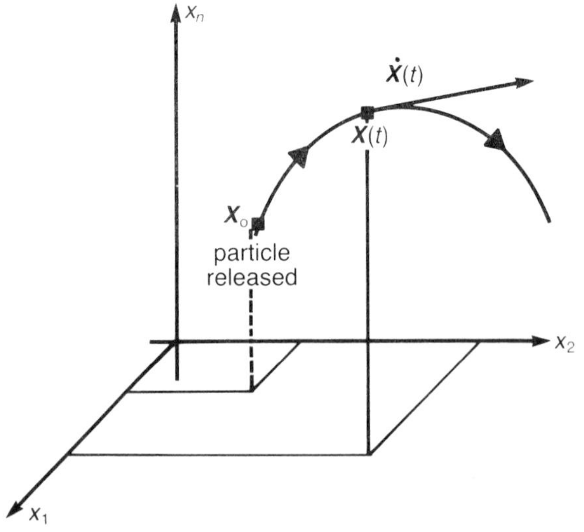

Fig. 10.1(b): A solution $\mathbf{X}(t)$

If a particle is released at position $\mathbf{X}_0 = \mathbf{X}(0)$ and $\mathbf{X}(t)$ specifies the position of the particle at time t, then the tangent vector $\dot{\mathbf{X}}(t)$ is also the velocity vector and so it must coincide with the wind velocity $\mathbf{f}(\mathbf{X}(t), t)$. For this reason, solutions to $\dot{\mathbf{X}} = \mathbf{f}(\mathbf{X}, t)$ are often called *trajectories* and positions $\mathbf{X}_1$ where the wind does not blow are called *stationary points.*

Definition 10.1 A point $\mathbf{X}_1$ in $\mathbf{R}^n$ is called a *stationary* or *critical point* in case $\mathbf{f}(\mathbf{X}_1, t) = 0$ for all t. If $\mathbf{f}(\mathbf{X}, t)$ can be written as a function of $\mathbf{X}$ alone, we call the system *autonomous*. Finally, if $\mathbf{f}(\mathbf{X}, t) = \mathbf{A}\,\mathbf{X}$ for some matrix $\mathbf{A}$, the system is called a *linear autonomous system.*

In the case of an autonomous system $\dot{\mathbf{X}} = \mathbf{f}(\mathbf{X})$, the vector field is independent of time and a solution $\mathbf{X} = \mathbf{X}(t)$ will have tangent vector $\dot{\mathbf{X}}(t)$ coinciding with the fixed velocity vector $\mathbf{f}(\mathbf{X}(t))$. These new concepts are illustrated in the following examples.

Example 10.1 For the differential equation $\dot{\mathbf{X}} = \mathbf{A}\,\mathbf{X}$ with

$$\mathbf{A} = \begin{bmatrix} -0.03 & 0.03 \\ 0.03 & -0.03 \end{bmatrix}$$

the critical points $\mathbf{X} = [x_1, x_2]$ must satisfy $\mathbf{AX} = \mathbf{0}$. Hence, $x_1 = x_2$ and we have an entire line of critical points as shown in Fig. 10.2.

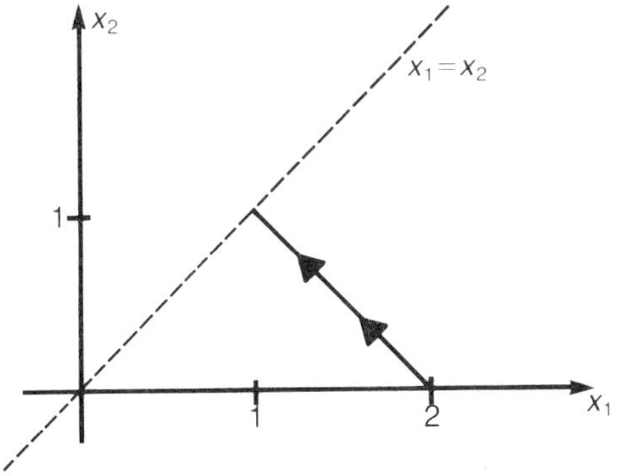

Fig. 10.2: The solution with $\mathbf{X}(0) = [2, 0]$

The solution satisfying $\mathbf{X}(0) = [2, 0]$, obtained by the methods of Chapter 2, is $\mathbf{X}(t) = [1, 1] + [1, -1]\ e^{-0.06t}$ and is shown in the figure. As $t \to +\infty$, $\mathbf{X}(t)$ approaches the stationary point $[1, 1]$.

Example 10.2 (Volterra predator–prey model) When x_1 = number of predators and x_2 = number of prey, the classical Volterra predator–prey model leads to the system of differential equations

$$\dot{x}_1 = -a\,x_1 + b\,x_1x_2$$

$$\dot{x}_2 = -c\,x_1x_2 + d\,x_2$$

where $a, b, c, d > 0$.

This system is autonomous and non-linear. Note that in the absence of predators (i.e. $x_1 = 0$), $\dot{x}_2 = dx_2$ and so the prey increase exponentially. In the absence of prey, $\dot{x}_1 = -ax_1$ and so the predators starve. The term $-c\,x_1x_2$ represents the death rate due to predators. The model assumes that this death rate is directly proportional to the number of possible encounters x_1x_2 between predator and prey at a particular time t.

To find stationary points, we set $\mathbf{f}(\mathbf{X}) = \mathbf{0}$ to obtain $x_1(-a + bx_2) = 0$ and $x_2(-cx_1 + d) = 0$. If $x_1 = 0$, then $dx_2 = 0$ or $x_2 = 0$. If $x_2 = a/b$, then $x_2 > 0$. From the second equation it follows that $x_1 = d/c$. Hence the stationary points are $[0, 0]$ and $[d/c, a/b]$.

Let $a = 0.1$, $d = 0.2$, $b = 0.002$, and $c = 0.0025$. The two critical points are then $[0, 0]$ and $[80, 50]$. Shown in Fig. 10.3 is the solution satisfying $\mathbf{X}(0) = [40, 40]$. Note that since the system is *non-linear*, none of our prior solution techniques apply. The solution in the figure was generated by applying one of the *numerical approximation methods* that we discuss later in the chapter. Note that the solution cycles and does not approach either stationary point. This is characteristic of all solution curves in the model (see [1], pages 586–589).

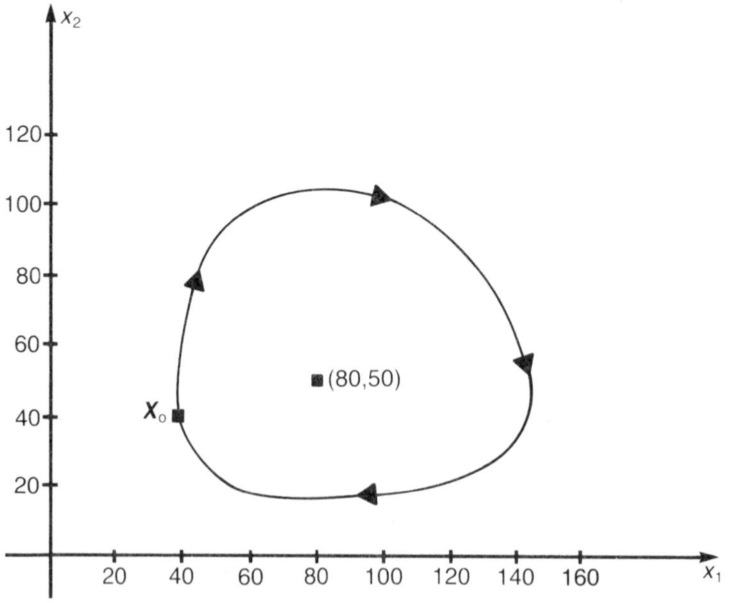

Fig. 10.3: The Lotka–Volterra cycle with $\mathbf{X}(0) = [40, 40]$

Example 10.3 As illustrated in Fig. 10.4, a drug is infused into the bloodstream at the known rate if I mg/minute and is *actively removed* from the body by an organ at the rate of $R(t)$ mg/minute.

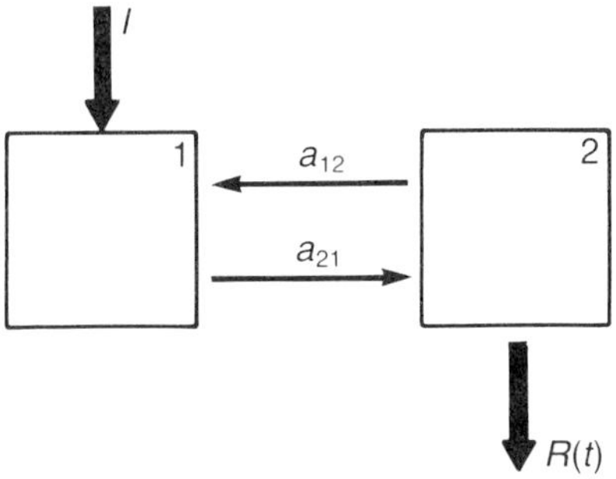

Fig. 10.4: The removal rate is given by a Michaelis–Menten function

This removal rate is governed by the Michaelis–Menten law

$$R(t) = \frac{ac_2(t)}{c_2(t) + b}$$

where $a, b > 0$. The system of differential equations then takes the form

$$\dot{x}_1 = -a_{21}x_1 + a_{12}x_2 + I$$

$$\dot{x}_2 = a_{21}x_1 - a_{12}x_2 - \frac{ac_2}{c_2 + b}$$

To find the critical points, set $\dot{\mathbf{X}} = \mathbf{0}$ and add the two equations to obtain $ac_2/(c_2 + b) = I$. Hence $c_2 = bI/(a - I)$ or $x_2 = bIV_2/(a - I)$. Using the first differential equation, we obtain

$$x_1 = \frac{-I^2 + aI + a_{12}bIV_2}{a_{21}(a - I)}$$

Note that when $I = 0$, the critical point reduces to $[0, 0]$ as expected. Also note that the critical point lies in the first quadrant if and only if $a > I$.

10.2 LINEARIZATION AND LOCAL STABILITY

Suppose that $\mathbf{X}_1$ is a stationary point for an autonomous system $\dot{\mathbf{X}} = \mathbf{f}(\mathbf{X})$. If $\partial f_i/\partial x_j$ exists and is continuous in a neighborhood of $\mathbf{X}_1$ for each i and j, then

$$\mathbf{f}(\mathbf{X}) = \mathbf{f}(\mathbf{X}_1) + \mathbf{f}'(\mathbf{X}_1)(\mathbf{X} - \mathbf{X}_1) + \epsilon(\mathbf{X} - \mathbf{X}_1)$$

where $|\epsilon(\mathbf{X} - \mathbf{X}_1)|/|\mathbf{X} - \mathbf{X}_1| \to 0$ as $\mathbf{X} \to \mathbf{X}_1$. Here $\mathbf{f}'(\mathbf{X}_1)$ is the *Jacobian matrix* $[\partial f_i/\partial x_j]$ evaluated at $\mathbf{X}_1$. Since $\mathbf{f}(\mathbf{X}_1) = \mathbf{0}$, we have, for $\mathbf{X}$ close to $\mathbf{X}_1$, the approximation

$$\dot{\mathbf{X}} = \mathbf{f}(\mathbf{X}) \approx \mathbf{f}'(\mathbf{X}_1)(\mathbf{X} - \mathbf{X}_1).$$

Letting $\mathbf{H} = \mathbf{X} - \mathbf{X}_1$, it follows that $\dot{\mathbf{H}} \approx \mathbf{f}'(\mathbf{X}_1)\,\mathbf{H}$. The linear system $\dot{\mathbf{H}} = \mathbf{f}'(\mathbf{X}_1)\,\mathbf{H}$ is called the *linearization of* $\dot{\mathbf{X}} = \mathbf{f}(\mathbf{X})$ *in a neighborhood of* $\mathbf{X}_1$. As we shall see, the behavior of solutions to the non-linear autonomous system $\dot{\mathbf{X}} = \mathbf{f}(\mathbf{X})$ near $\mathbf{X}_1$ can often be analyzed by finding the eigenvalues of the Jacobian matrix $\mathbf{f}'(\mathbf{X}_1)$.

Imagine that a particle has been displaced from a stationary point $\mathbf{X}_1$ to a nearby point $\mathbf{X}_0$. Will the particle return to $\mathbf{X}_1$? To determine when this occurs, we must investigate the solution $\mathbf{X}(t)$ with $\mathbf{X}(0) = \mathbf{X}_0$ and determine whether $\operatorname{limit}_{t \to +\infty} \mathbf{X}(t) = \mathbf{X}_1$. The precise definition of a *locally stable* critical point is given below.

Definition 10.2 A stationary point $\mathbf{X}_1$ for an autonomous system $\dot{\mathbf{X}} = \mathbf{f}(\mathbf{X})$ is called *locally stable* (the term 'asymptotically locally stable' is also used) if and only if for each $\epsilon > 0$ there is a corresponding $\delta > 0$ such that if $|\mathbf{X}_0 - \mathbf{X}_1| < \delta$, then

(1) $|\mathbf{X}(t) - \mathbf{X}_1| < \epsilon$ for all $t > 0$

(2) $\operatorname{limit}_{t \to +\infty} \mathbf{X}(t) = \mathbf{X}_1$

where $\mathbf{X}(t)$ is the solution to $\dot{\mathbf{X}} = \mathbf{f}(\mathbf{X})$ satisfying $\mathbf{X}(0) = \mathbf{X}_0$.

The definition of a locally stable critical point is illustrated in Fig. 10.5. For any given neighborhood of $\mathbf{X}_1$, the solution $\mathbf{X}(t)$ will remain in the neighborhood and eventually approach $\mathbf{X}_1$ provided $\mathbf{X}_0 = \mathbf{X}(0)$ is chosen sufficiently close to $\mathbf{X}_1$.

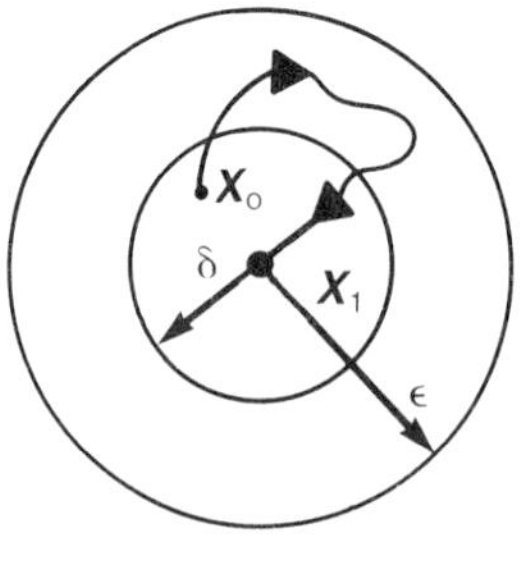

Fig. 10.5: Local stability

If $\dot{\mathbf{X}} = \mathbf{A}\,\mathbf{X}$ and $\mathbf{A}$ has spectral decomposition $\sum_{i=1}^{n} \lambda_i \mathbf{Z}_i$, then solutions are of the form

$$\mathbf{X}(t) = e^{\mathbf{A}t}\,\mathbf{X}_0 = \sum_{i=1}^{n} e^{\lambda_i t}\,\mathbf{Z}_i \mathbf{X}_0$$

If $\lambda_k = \mu_k + i\nu_k$, then

$$|\mathbf{X}(t)| \leqslant \sum_{i=1}^{n} e^{\mu_i t} \, |\mathbf{Z}_i \, \mathbf{X}_0|.$$

We may conclude from this inequality that if $\mu_i < 0$ for each i then $\mathbf{X}(t)$ will always return to the stationary point $\mathbf{X}_1 = \mathbf{0}$. If, however, one of the eigenvalues, say λ_1, has $\mu_1 \geqslant 0$, let $\mathbf{E}_1$ be a non-zero eigenvector corresponding to λ_1 and let $\mathbf{X}(t)$ be the solution

$$\mathbf{X}(t) = \frac{1}{n} \mathbf{E}_1 \, e^{\lambda_1 t}$$

Then $\mathbf{X}(0) = (1/n)\,\mathbf{E}_1$ can be made arbitrarily close to $\mathbf{0}$. But note that $|\mathbf{X}(t)| = (1/n)\,|\mathbf{E}_1|\, e^{\mu_1 t} \geqslant (1/n)\,|\mathbf{E}_1|$ and so $\mathbf{X}(t)$ does not return to $\mathbf{X}_1 = \mathbf{0}$. A similar argument can be presented using Theorem 2.5 when $\mathbf{A}$ does not possess n linearly independent eigenvectors. We have therefore established the following theorem.

Theorem 10.1 $\mathbf{X}_1 = 0$ is a locally stable critical point for $\dot{\mathbf{X}} = \mathbf{A}\,\mathbf{X}$ if and only if all eigenvalues of $\mathbf{A}$ have negative real part.

Example 10.4 For the linear system in Example 10.1, the eigenvalues are 0 and -0.06 and solutions are of the form

$$\mathbf{X}(t) = c_1\,[1, 1] + c_2\,[1, -1]\; e^{-0.06t}$$

According to Theorem 10.1, $\mathbf{X}_1 = \mathbf{0}$ is not a locally stable critical point. In fact, none of the critical points on the line $x_1 = x_2$ are locally stable. This can be seen from the solution curves sketched in Fig. 10.6.

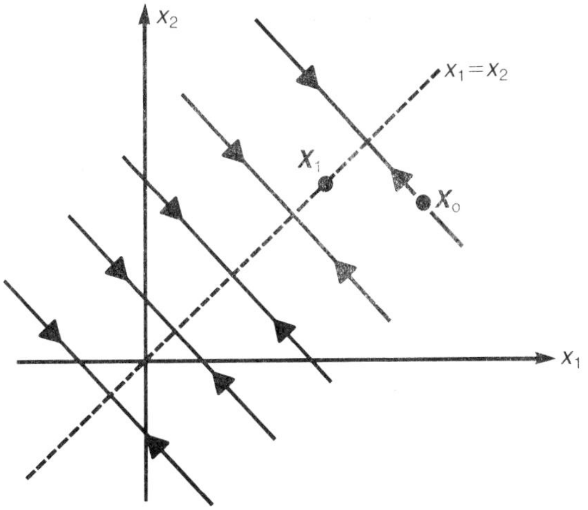

Fig. 10.6: Solutions for example 10.4

When displaced slightly to the right or left from $\mathbf{X}_1$, $\mathbf{X}_0$ will not return to $\mathbf{X}_1$.

If $\mathbf{X}_1$ is a critical point for the non-linear autonomous system $\dot{\mathbf{X}} = \mathbf{f}(\mathbf{X})$, then the linearization near $\mathbf{X}_1$ is $\dot{\mathbf{H}} = \mathbf{A}\,\mathbf{H}$, where $\mathbf{H} = \mathbf{X} - \mathbf{X}_1$ and $\mathbf{A} = \mathbf{f}'(\mathbf{X}_1) = [\partial f_i/\partial x_j(\mathbf{X}_1)]$. For the linear system, $\mathbf{H}(t) \to \mathbf{0}$, that is, $\mathbf{X}(t) \to \mathbf{X}_1$, when all eigenvalues of the Jacobian matrix $\mathbf{A} = \mathbf{f}'(\mathbf{X}_1)$ have negative real part. Our next theorem asserts that solutions to the original system show the same behavior.

Theorem 10.2 Let $\mathbf{X}_1$ be a critical point of the autonomous system $\dot{\mathbf{X}} = \mathbf{f}(\mathbf{X})$. Then $\mathbf{X}_1$ is locally stable if and only if all eigenvalues of the Jacobian matrix $\mathbf{f}'(\mathbf{X}_1)$ have negative real part.

A strict proof of this theorem requires some technical inequalities we have not developed and so the proof is omitted. Note that Theorem 10.2 does not tell us how close $\mathbf{X}_0$ must be to $\mathbf{X}_1$ in order to return. This is illustrated in the following example.

Example 10.5 The system of differential equations

$$\dot{x}_1 = 0.004\,x_1(50 - x_1 - 0.75\,x_2)$$

$$\dot{x}_2 = 0.001\,x_2(100 - x_2 - 3x_1)$$

is a special case of the famous *Lotka–Volterra competition model* in which two species compete for a common resource. Note that when $x_2 = 0$, $\dot{x}_1 = 0.004\,x_1(50 - x_1)$ and so the first population grows logistically to the limiting value of 50. Likewise when $x_1 = 0$ the second population grows logistically. When x_1 and $x_2 > 0$, each population inhibits the growth of the other at a rate proportional to the encounter rate, which is itself proportional to x_1x_2 at a particular time t.

To find stationary points, note that when $x_1 = 0$, we have $x_2(100 - x_2) = 0$. Therefore [0, 0] and [0, 100] are two such points. Also, when $x_2 = 0$, $x_1 = 0$ or 50. A third stationary point is then [50, 0]. The final stationary point is the solution [20, 40] to the system

$$\begin{aligned} x_1 + 0.75\,x_2 &= 50 \\ 3x_1 + x_2 &= 100 \end{aligned}$$

To investigate local stability, we examine the eigenvalues of the Jacobian matrix

$$\mathbf{f}'(\mathbf{X}) = \begin{bmatrix} 0.2 - 0.008x_1 - 0.003x_2 & -0.003x_1 \\ -0.003x_2 & 0.1 - 0.002x_2 - 0.003x_1 \end{bmatrix}$$

For $\mathbf{X}_1 = [0, 0]$, the eigenvalues are 0.2 and 0.1 and so this critical point is not locally stable. For $\mathbf{X}_1 = [0, 100]$,

$$\mathbf{f}'(\mathbf{X}_1) = \begin{bmatrix} -0.1 & 0 \\ -0.3 & -0.1 \end{bmatrix}$$

and so both eigenvalues are negative. Hence $\mathbf{X}_1 = [0, 100]$ is locally stable. Likewise $\mathbf{X}_1 = [50, 0]$ is locally stable. Finally, for $\mathbf{X}_1 = [20, 40]$,

$$\mathbf{f}'(\mathbf{X}_1) = \begin{bmatrix} -0.08 & -0.06 \\ -0.12 & -0.04 \end{bmatrix}$$

which has eigenvalues 0.0272 and −0.1472. This critical point is therefore unstable.

Note that none of our computations predict the long-range behavior of solutions $\mathbf{X}(t)$ for initial conditions such as $\mathbf{X}(0) = [40, 80]$ or $[80, 40]$. These initial values are not close to the critical points. Shown in Fig. 10.7, however, are the solution curves obtained by using a numerical approximation method. Also shown are two solutions near the unstable critical points.

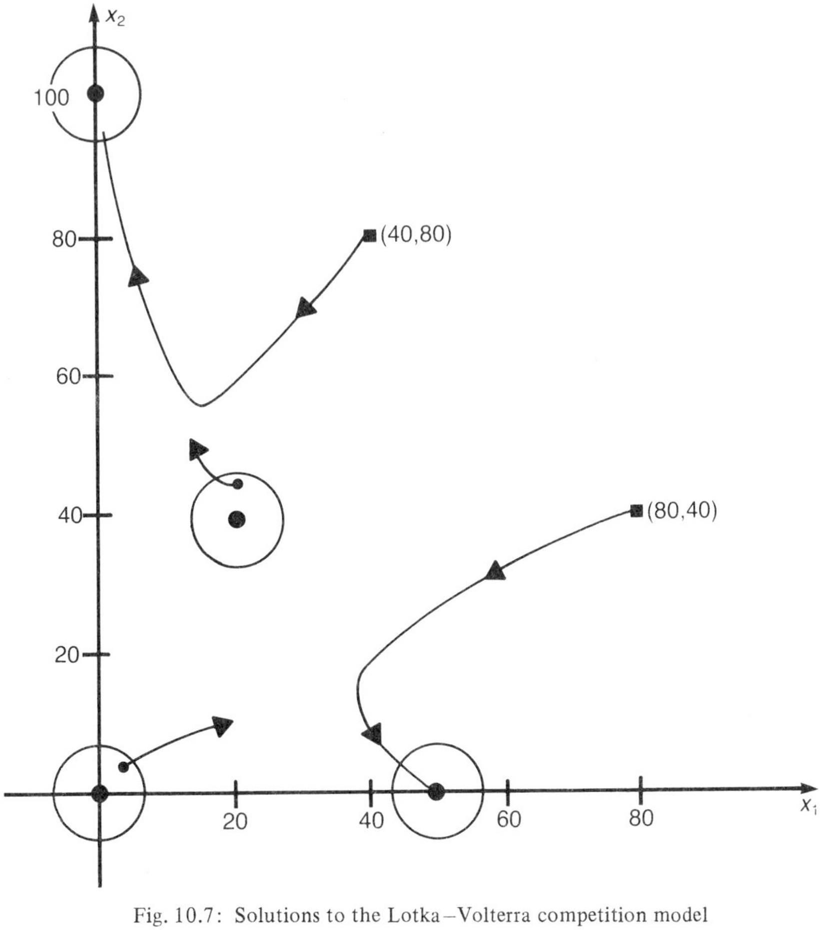

Fig. 10.7: Solutions to the Lotka–Volterra competition model

Many studies in biology involve the use of radioactive tracers and the assumption is made that the biological system processes the tracer in exactly the same manner as the normal non-radioactive substance. If we assume that the system is in equilibrium before the tracer is introduced, then the introduction of a *small amount* of tracer results in a perturbation or small displacement from equilibrium. Put in mathematical language, suppose that $\dot{\mathbf{X}} = \mathbf{f}(\mathbf{X})$ and $\mathbf{f}(\mathbf{X}_1) = \mathbf{0}$. If the vector $\mathbf{H}$ specifies the amount of radioactive tracer in the system, then $\mathbf{X} = \mathbf{X}_1 + \mathbf{H}$ and

$$\dot{\mathbf{H}} = \dot{\mathbf{X}} \approx \mathbf{f}'(\mathbf{X}_1)\,\mathbf{H}$$

when $|\mathbf{H}|$ is small. Thus the dynamics of the tracer can be approximated by the *linear system* $\dot{\mathbf{H}} = \mathbf{A}\,\mathbf{H}$ with $\mathbf{A} = \mathbf{f}'(\mathbf{X}_1)$ when the critical point $\mathbf{X}_1$ is locally stable. This argument is often given to justify the use of linear compartment models in tracer studies. Note, however, that we have no guarantee that the Jacobian matrix $\mathbf{A} = \mathbf{f}'(\mathbf{X}_1)$ will have the special form of a transfer matrix.

10.3 NUMERICAL METHODS

When the system of differential equations $\dot{\mathbf{X}} = \mathbf{f}(\mathbf{X}, t)$ is non-linear, rarely can we find explicit expressions for solutions $\mathbf{X}(t)$. To close this chapter, we will discuss three of the more popular numerical algorithms that can be used to generate approximate solutions $\bar{\mathbf{X}}(t)$.

If the vector-valued function $\mathbf{X}(t)$ is $(n + 1)$-times continuously differentiable in a neighborhood of t, then

$$\mathbf{X}(t+h) = \mathbf{X}(t) + h\,\mathbf{X}'(t) + \frac{h^2}{2!}\mathbf{X}''(t) + \ldots + \frac{h^n}{n!}\mathbf{X}^{(n)}(t) + \epsilon$$

where $\epsilon = 0(h^{n+1})$, that is, $|\epsilon| \leqslant c\,h^{n+1}$ for some constant c. If $\mathbf{X}(t)$ is a solution of $\dot{\mathbf{X}} = \mathbf{f}(\mathbf{X}, t)$, then

$$\mathbf{X}(t+h) = \mathbf{X}(t) + h\,\mathbf{X}'(t) + \epsilon = \mathbf{X}(t) + h\,\mathbf{f}(\mathbf{X}(t), t) + \epsilon$$

where $\epsilon = 0(h^2)$. We may therefore generate approximations to the solution satisfying $\mathbf{X}(0) = \mathbf{X}_0$ by employing the following recursion:

> *Euler's method*:
>
> $$\bar{\mathbf{X}}(t+h) = \bar{\mathbf{X}}(t) + h\,\mathbf{f}(\bar{\mathbf{X}}(t), t)$$
>
> with $\bar{\mathbf{X}}(0) = \mathbf{X}_0$

Note that in the special case $\dot{\mathbf{X}} = \mathbf{A}\,\mathbf{X}$, the formula becomes $\bar{\mathbf{X}}(t+h) = (\mathbf{I} + h\mathbf{A})\,\bar{\mathbf{X}}(t)$. Since the true solution satisfies $\mathbf{X}(t+h) = e^{\mathbf{A}h}\,\mathbf{X}(t)$, we are using the first two terms in the series expansion for $e^{\mathbf{A}h}$ in our approximation. In general, Euler's method will produce estimates for $\mathbf{X}(h)$, $\mathbf{X}(2h), \ldots, \mathbf{X}(nh), \ldots,$ as illustrated in the following example.

Example 10.6 If we apply Euler's method to the system in Example 10.1, then

$$\bar{\mathbf{X}}(t+h) = \begin{bmatrix} 1-0.03h & 0.03h \\ 0.03h & 1-0.03h \end{bmatrix} \bar{\mathbf{X}}(t)$$

with $\bar{\mathbf{X}}(0) = [2, 0]$. Shown in Table 10.1 are the approximations with $h = 1$ and 0.1, together with the actual solution obtained in Example 10.1.

Table 10.1

Time t	$\bar{\mathbf{X}}(t)$ with $h = 1$	$\bar{\mathbf{X}}(t)$ with $h = 0.1$	$\mathbf{X}(t)$
0	[2, 0]	[2, 0]	[2, 0]
1	[1.94, 0.06]	[1.94159, 0.05841]	[1.94176, 0.05824]
2	[1.8836, 0.1164]	[1.8866, 0.1134]	[1.88692, 0.11308]
5	[1.7339, 0.26610]	[1.74015, 0.25985]	[1.74082, 0.25918]
10	[1.5386, 0.46138]	[1.5478, 0.45218]	[1.54881, 0.451188]
20	[1.2901, 0.70989]	[1.3001, 0.69989]	[1.30119, 0.698806]
30	[1.1563, 0.843745]	[1.1644, 0.83559]	[1.1653, 0.83470]
40	[1.0842, 0.91584]	[1.09006, 0.90993]	[1.09072, 0.909282]
50	[1.0453, 0.95467]	[1.04933, 0.95065]	[1.04979, 0.95021]
100	[1.0021, 0.997948]	[1.00242, 0.997546]	[1.00248, 0.997521]

Notice that the better approximation for $\mathbf{X}(t)$ was obtained using the smaller value of h. This is usually the case.

To obtain higher accuracy for a given value of h, we next work with the first three terms $\mathbf{X}(t) + h\,\mathbf{X}'(t) + (h^2/2)\,\mathbf{X}''(t)$ in the Taylor expansion for $\mathbf{X}(t+h)$. Although $\mathbf{X}'(t) = \mathbf{f}(\mathbf{X}(t), t)$, the computation of $\mathbf{X}''(t)$ from the differential equation $\dot{\mathbf{X}} = \mathbf{f}(\mathbf{X}(t), t)$ will involve the partial derivatives of $\mathbf{f}$. In order to avoid working with these partials, we do some further approximations:

$$\mathbf{X}''(t) \approx \frac{\mathbf{X}'(t+h) - \mathbf{X}'(t)}{h}$$

$$= \frac{1}{h}\,[\mathbf{f}(\mathbf{X}(t+h), t+h) - \mathbf{f}(\mathbf{X}(t), t)]$$

Hence

$$\mathbf{X}(t+h) \approx \mathbf{X}(t) + h\,\mathbf{f}(\mathbf{X}(t), t) + \frac{h^2}{2}\,\frac{1}{h}\,[\mathbf{f}(\mathbf{X}(t+h), t+h) - \mathbf{f}(\mathbf{X}(t), t)]$$

$$\approx \mathbf{X}(t) + \frac{h}{2}\,\mathbf{f}(\mathbf{X}(t), t) + \frac{h}{2}\,\mathbf{f}(\mathbf{X}(t) + h\,\mathbf{f}(\mathbf{X}(t), t), t+h)$$

where we have used the Euler formula approximation for $\mathbf{X}(t+h)$ on the right-hand side. It can be shown that the error ϵ is $0(h^3)$. This approximation is the basis for the 2nd order Runge–Kutta method:

2nd order Runge–Kutta method:

$$\bar{\mathbf{X}}(t+h) = \bar{\mathbf{X}}(t) + \frac{h}{2}(\mathbf{K}_1 + \mathbf{K}_2)$$

where $\mathbf{K}_1 = \mathbf{f}(\bar{\mathbf{X}}(t), t)$ and $\mathbf{K}_2 = \mathbf{f}(\bar{\mathbf{X}}(t) + h\,\mathbf{K}_1, t+h)$

When $\mathbf{f}(\mathbf{X}, t) = \mathbf{A}\,\mathbf{X}$, it is not hard to show that the recursion $\bar{\mathbf{X}}(t+h) = \bar{\mathbf{X}}(t) + (h/2)(\mathbf{K}_1 + \mathbf{K}_2)$ reduces to $\bar{\mathbf{X}}(t+h) = (\mathbf{I} + h\mathbf{A} + (h^2/2\,!)\,\mathbf{A}^2)\,\bar{\mathbf{X}}(t)$.

Example 10.7 Approximate the solution to the competition model in Example 10.5 that satisfies the initial condition $\mathbf{X}(0) = [20, 30]$.

Solution 10.7 We will first approximate the solution using the 2nd order Runge–Kutta algorithm with $h = 1$ and then use $h = 0.5$ to check the accuracy. For the system in question, $\dot{\mathbf{X}} = \mathbf{f}(\mathbf{X})$ with

$$\mathbf{f}(\mathbf{X}) = [0.004\,x_1(50 - x_1 - 0.75\,x_2), 0.001\,x_2(100 - x_2 - 3x_1)]$$

and so $\mathbf{K}_1 = \mathbf{f}(\mathbf{X})$ and $\mathbf{K}_2 = \mathbf{f}(\mathbf{X} + h\,\mathbf{K}_1)$. Shown in Table 10.2 are the approximations corresponding to $h = 1$ and $h = 0.5$ respectively. As you can see, there is little difference between the two approximations when both are rounded off to two correct decimal places.

Table 10.2

Time t	$\bar{\mathbf{X}}(t)$ with $h = 1$	$\bar{\mathbf{X}}(t)$ with $h = 0.5$
0	[20, 30]	[20, 30]
1	[20.575, 30.2697]	[20.576, 30.2705]
2	[21.1047, 30.484]	[21.1066, 30.4854]
5	[22.4525, 30.8387]	[22.4561, 30.8412]
10	[24.0991, 30.7107]	[24.1033, 30.7131]
20	[26.3592, 28.9718]	[26.3618, 28.9722]
50	[33.3798, 19.354]	[33.3811, 19.3521]
100	[46.1868, 3.95408]	[46.1883, 3.95233]
150	[49.6234, 0.37802]	[49.6238, 0.377575]
200	[49.9685, 0.03150]	[49.9686, 0.03146]

Perhaps the most popular of all numerical methods for approximating solutions to differential equations is the *4th order Runge–Kutta method*. Briefly, the method is based on the approximation

$$\mathbf{X}(t+h) \approx \mathbf{X}(t) + h\,\mathbf{X}'(t) + \frac{h^2}{2!}\mathbf{X}''(t) + \frac{h^3}{3!}\mathbf{X}^{(3)}(t) + \frac{h^4}{4!}\mathbf{X}^{(4)}(t)$$

but the algorithm is set up so that no higher order partials of **f** are needed in the computation:

4th order Runge–Kutta method:

$$\bar{\mathbf{X}}(t+h) = \bar{\mathbf{X}}(t) + \frac{h}{6}(\mathbf{K}_1 + 2\mathbf{K}_2 + 2\mathbf{K}_3 + \mathbf{K}_4)$$

where $\mathbf{K}_1 = \mathbf{f}(\bar{\mathbf{X}}(t), t)$, $\mathbf{K}_2 = \mathbf{f}\left(\bar{\mathbf{X}}(t) + \frac{h}{2}\mathbf{K}_1,\, t + \frac{h}{2}\right)$

$\mathbf{K}_3 = \mathbf{f}\left(\bar{\mathbf{X}}(t) + \frac{h}{2}\mathbf{K}_2,\, t + \frac{h}{2}\right)$ and $\mathbf{K}_4 = \mathbf{f}(\bar{\mathbf{X}}(t) + h\mathbf{K}_3,\, t + h)$

It can be shown that the error at each step is $0(h^5)$ and that, for the special case in which $\mathbf{f}(\mathbf{X}(t), t) = \mathbf{A}\,\mathbf{X}$, the recursion reduces to

$$\bar{\mathbf{X}}(t+h) = \left(\mathbf{I} + \mathbf{A}h + \frac{\mathbf{A}^2 h^2}{2!} + \frac{\mathbf{A}^3 h^3}{3!} + \frac{\mathbf{A}^4 h^4}{4!}\right)\bar{\mathbf{X}}(t)$$

We will use this new method to approximate the solution to the non-linear predator–prey model presented in Example 10.2

Example 10.8 Approximate the solution to the non-linear autonomous system

$$\dot{x}_1 = x_1(-.1 + 0.002\,x_2)$$

$$\dot{x}_2 = x_2(-0.0025\,x_1 + 0.2)$$

with $\mathbf{X}(0) = [40, 40]$.

Solution 10.8 For autonomous sytems, the Runge–Kutta algorithm simplifies with $\mathbf{K}_1 = \mathbf{f}(\mathbf{X})$, $\mathbf{K}_2 = \mathbf{f}(\mathbf{X} + (h/2)\,\mathbf{K}_1)$, $\mathbf{K}_3 = \mathbf{f}(\mathbf{X} + (h/2)\,\mathbf{K}_2)$ and $\mathbf{K}_4 = \mathbf{f}(\mathbf{X} + h\,\mathbf{K}_3)$, where $\mathbf{f}(\mathbf{X}) = [x_1(-0.1 + 0.002\,x_2), x_2(-0.0025\,x_1 + 0.2)]$. Shown in Table 10.3 is the approximation corresponding to $h = 1$. Subsequent approximations with $h = 0.05$ and $h = 0.01$ produce essentially the same results. The solution $\mathbf{X}(t)$ is periodic with period $p \approx 47$.

Table 10.3

Time t	0	2	4	6
$\bar{\mathbf{X}}(t)$	[40, 40]	[39.1, 49.0]	[39.8, 60.05]	[42.5, 73.0]
Time t	8	10	12	14
$\bar{\mathbf{X}}(t)$	[47.9, 87.0]	[57.0, 100.0]	[71.0, 108.6]	[90.1, 108.4]
Time t	16	18	20	22
$\bar{\mathbf{X}}(t)$	[111.8, 97.7]	[130.6, 79.3]	[141.1, 59.7]	[141.9, 43.8]
Time t	24	26	28	30
$\bar{\mathbf{X}}(t)$	[135.1, 32.6]	[124.1, 25.4]	[111.4, 21.0]	[98.7, 18.6]
Time t	32	34	36	38
$\bar{\mathbf{X}}(t)$	[86.8, 17.4]	[76.1, 17.3]	[66.9, 18.1]	[59.0, 19.7]
Time t	40	42	44	46
$\bar{\mathbf{X}}(t)$	[52.5, 22.2]	[47.3, 25.9]	[43.4, 30.8]	[40.7, 37.2]
Time t	47	48		
$\bar{\mathbf{X}}(t)$	[39.8, 41.1]	[39.3, 45.5]		

The Runge–Kutta methods are very useful in working with linear non-homogeneous systems

$$\dot{\mathbf{X}} = \mathbf{A}\,\mathbf{X} + \mathbf{I}(t)$$

when the driving function does not fit into one of the simple forms given in Chapter 7. For example, in constructing an ecosystem model in which the sun drives the system and supplies light for photosynthesis, we might use a periodic driving function $\mathbf{I}(t)$, where $\mathbf{I}(t + 24) = \mathbf{I}(t)$ and

$$\mathbf{I}(t) = \begin{cases} a + b \sin \omega t, & 0 \leqslant t \leqslant D \\ 0, & D \leqslant t \leqslant 24 \end{cases}$$

(Time $t = 0$ corresponds to sunrise and $t = D$ corresponds to sunset.) It is difficult to construct a particular solution using the methods of Chapter 7. If we select the step size h so that $nh = D$ for some n, then a Runge–Kutta method can be applied to give good results. Of course, all transfer coefficients and parameters must be found before the method can be invoked. As we shall see in Chapter 11, this usually is the case in constructing ecosystem models.

EXERCISES

Programming Exercise

The program requested in Exercise 1 will prove useful in doing many of the exercises that follow.

1 Write a program that solves a system of differential equations $\dot{\mathbf{X}} = \mathbf{f}(\mathbf{X}, t)$

subject to $\mathbf{X}(0) = \mathbf{X}_0$, where $\mathbf{X} = [x_1, x_2, x_3]$ with the following features.

(a) The user enters the dynamics $\mathbf{f}(\mathbf{X}, t) = [f_1(\mathbf{X}, t), f_2(\mathbf{X}, t), f_3(\mathbf{X}, t)]$.
(b) User is allowed to select either the Euler, 2nd order Runge–Kutta, or 4th order Runge–Kutta method to generate solutions.
(c) User selects the step size and solution interval $[0, T]$.
(d) Solution approximations are printed out every k steps (i.e. at times $t = kh$, $2kh, \ldots$)

Part A

For each of the following systems of differential equations, (a) find all critical points, (b) analyze each critical point for local stability, (c) find the solution satisfying the given initial condition, and (d) compute (if possible) $\lim_{t\to\infty} \mathbf{X}(t)$.

2
$$\dot{x}_1 = -2x_1 + 3x_2$$
$$\dot{x}_2 = x_1 - 3x_2$$
with $\mathbf{X}(0) = [4, 3]$

3
$$\dot{x}_1 = -2x_1 + 3x_2 + x_3$$
$$\dot{x}_2 = x_1 - 3x_2 + 2x_3$$
$$\dot{x}_3 = x_1 - 3x_3$$
with $\mathbf{X}(0) = [5, 5, 5]$

4
$$\dot{x}_1 = 0.01\, x_1(100 - x_1 - x_2)$$
$$\dot{x}_2 = \frac{1}{120} x_2(60 - x_2 - 0.2x_1)$$
with $\mathbf{X}(0) = [30, 80]$

5
$$\dot{x}_1 = 0.1x_1 - 0.002\, x_1x_2$$
$$\dot{x}_2 = -0.001x_1x_2 + 0.0025\, x_2(100 - x_2)$$
with $\mathbf{X}(0) = [100, 100]$

6
$$\dot{x}_1 = x_1^2 + x_2^2 - 6$$
$$\dot{x}_2 = x_1^2 - x_2 \qquad \text{with } \mathbf{X}(0) = [4, 4]$$

7
$$\dot{x}_1 = -0.04\, x_1 + 15x_1\, e^{-(x_1 + x_2)}$$
$$\dot{x}_2 = 0.05x_1 - 0.1x_2 \qquad \text{with } \mathbf{X}(0) = [10, 10]$$

8
$$\dot{x}_1 = -2x_1 + x_2 + 10$$
$$\dot{x}_2 = 2x_1 - x_2 - \frac{15x_2}{x_2 + 5} \qquad \text{with } \mathbf{X}(0) = [0, 0]$$

9
$$\dot{x}_1 = -0.05x_1 + 0.08x_2 + 0.03x_3 + 20$$
$$\dot{x}_2 = 0.01x_1 - 0.1x_2 + 0.05x_3$$
$$\dot{x}_3 = 0.02x_1 + 0.02x_2 - 0.15x_3 \qquad \text{with } \mathbf{X}(0) = [0, 0, 0]$$

10 Examine the critical points in Example 10.2 for local stability.

11 Determine conditions under which the critical point in Example 10.3 is locally stable.

12 The Lotka–Volterra competition model for two species leads to the system of differential equations

$$\dot{x}_1 = \frac{r_1}{K_1} x_1(K_1 - x_1 - \alpha_{21}x_2)$$

$$\dot{x}_2 = \frac{r_2}{K_2} x_2(K_2 - x_2 - \alpha_{12}x_1)$$

where $r_1, r_2, K_1, K_2, \alpha_{12}, \alpha_{21} > 0$.

(a) $x_2 \equiv 0$, solve the first equation for x_1.

(b) Show that if $\alpha_{21}\alpha_{12} \neq 1$, then there are two critical points $\mathbf{X} = \mathbf{0}$ and $\mathbf{X} = \mathbf{X}_1$. When is $\mathbf{X}_1$ in the first quadrant?

(c) Determine conditions on the parameters which will insure that $\mathbf{X}_1$ is locally stable. Is $\mathbf{X} = \mathbf{0}$ ever locally stable?

13 Find the period of the cyclical solution in Example 10.8 which satisfies the initial condition $\mathbf{X}(0) = [60, 60]$.

14 Under what conditions will a linear system $\dot{\mathbf{X}} = \mathbf{A}\,\mathbf{X} + \mathbf{F}(t)$ have critical points? What additional conditions are needed to guarantee that the critical points are locally stable?

15 Show that when $\mathbf{f}(\mathbf{X}, t) = \mathbf{A}\,\mathbf{X}$

(a) the 2nd order Runge–Kutta algorithm reduces to

$$\bar{\mathbf{X}}(t+h) = \left(\mathbf{I} + h\,\mathbf{A} + \frac{h^2}{2}\,\mathbf{A}^2\right)\bar{\mathbf{X}}(t)$$

(b) the 4th order Runge–Kutta algorithm reduces to

$$\bar{\mathbf{X}}(t+h) = \left(\mathbf{I} + h\,\mathbf{A} + \frac{h^2}{2}\,\mathbf{A}^2 + \frac{h^3}{6}\,\mathbf{A}^3 + \frac{h^4}{24}\,\mathbf{A}^4\right)\bar{\mathbf{X}}(t)$$

16 Estimate $\mathbf{X}(25)$ if $\mathbf{X}(t)$ satisfies the linear system $\dot{\mathbf{X}} = \mathbf{A}\,\mathbf{X} + \begin{bmatrix} F(t) \\ 0 \end{bmatrix}$ where $\mathbf{X}(0) = [0, 0]$,

$$\mathbf{A} = \begin{bmatrix} -2 & 1 \\ 1 & -1 \end{bmatrix}$$

and $F(t)$ takes the form

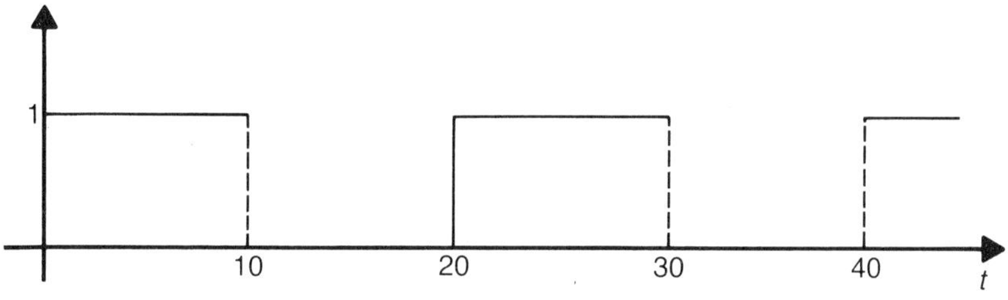

REFERENCES

[1] M. Braun, *Differential Equations and their Applications*, Springer-Verlag, 1975.

FURTHER READING

J. D. Murray, *Lectures on Nonlinear Differential Equations in Biology*, Clarendon Press (Oxford), 1977.

D. Sanchez, *Ordinary Differential Equations and Stability Theory: an Introduction*, W. H. Freeman (San Francisco), 1968.

CHAPTER 11

Applications to Systems Ecology

11.1 INTRODUCTION

The modeling of the transfer of some entity (e.g. radionuclide, nutrient, energy) through an ecosystem is a relatively recent phenomenon dating back to the early 1960s. The success of compartmental models in physiology and pharmacology, the advent of the computer, and the extremely important problem of assessing the effects of the products of radioactive fallout on ecosystems and food chains were the major factors that led to the creation of the field of *systems ecology*. In the 1960s and early 1970s, most of the models constructed were *linear compartment models* which led to differential equations of the form $\dot{\mathbf{X}} = \mathbf{A}\,\mathbf{X} + \mathbf{F}(t)$. In recent years, the majority of models tend to be highly non-linear with flux rates and driving functions constructed by fitting curves to the available data. Nevertheless, linear models can be useful and in a recent survey R. V. O'Neill [6] has identified several areas in which linear models are appropriate.

Linear models can often be used to study the cycling of a radionuclide through an ecosystem. Here the assumption is made that the non-radioactive counterpart has reached steady state in the various components of the system and that the introduction of the radionuclide results in a small displacement from equilibrium. As an example of such a model, we will present a study of Strontium-90 cycling in a Puerto Rican rain forest [4].

Frequently in modeling large scale ecosystems involving oceans and large land masses, there is either little data available or the known data are very crude. In such a case it makes little sense to construct a complex high resolution model. A simple linear model can serve as a *preliminary model* and is certainly preferable to mere guesswork on the behavior of the system. As an example of such a model, we will present a model of Hett and O'Neill [3] on the Aleutian Islands ecosystem.

11.2 ESTIMATION OF MODEL PARAMETERS

As we have seen in Chapters 8 and 9, in physiological tracer experiments, the transfer coefficients cannot be measured directly and remain unknowns in the model. The experiment takes place over a short enough time period that one can make partial observations of the system from which transfer coefficients, fluxes, and volumes can often be estimated. In ecological applications, the situation is quite different. First of all, the transport mechanisms are often very slow with the result that it would take years of data to perform curve fits similar to those in Chapter 9. Secondly, the matrices are almost always of large order. For the two ecosystems we will study in this chapter, the transfer matrices are 5×5 and 9×9 respectively. Finally, the purpose of an ecological model is not to estimate a_{ij}, r_{ij}, or V_j but to predict the state $\mathbf{X}(t)$ of the ecosystem years in the future. The systems ecologist must therefore produce estimates of the transfer coefficients and determine appropriate expressions for each driving function. This can require considerable experimental work in the field and, in the examples that follow, we will describe typical experiments and the appropriate mathematical formulas. Once an appropriate transfer matrix has been found, we can study the effects of various perturbations on the system by altering initial conditions and transfer coefficients.

In some cases, water will be the transport mechanism and, as a result of the flow, the tracer will be moved passively from one compartment to another. According to the formula developed on page 110,

$$\boxed{a_{ij} = F_{ij}/V_j}$$

This formula was used in Example 8.1 in a model for pollution in the Great Lakes. From estimates for the volume of Lake Erie and the net flow rate we obtained $a_{01} = F_{01}/V_1 \approx (175 \text{ km}^3/\text{yr})/458 \text{ km}^3 = 0.382/\text{year}$. In many cases, however, it is difficult to measure the water volume in a compartment and other methods are needed.

If we assume that the system has reached the *steady state* $\hat{\mathbf{X}} = [\hat{x}_1, \hat{x}_2, \ldots, \hat{x}_n]$, then the flux r_{ij} from compartment j to compartment i is $r_{ij} = a_{ij}\hat{x}_j$ and we obtain

$$\boxed{a_{ij} = r_{ij}/\hat{x}_j}$$

When water is the transport mechanism, the flux r_{ij} can be computed from the formula $r_{ij} = F_{ij} \hat{c}_j$, where $\hat{c}_j$ is the tracer concentration in the water flowing from compartment j to compartment i. Thus:

$$a_{ij} = \frac{F_{ij} \hat{c}_j}{\hat{x}_j}$$

These last two formulas are illustrated in the following examples.

Example 11.1 In the cycling of stable strontium through a forest ecosystem, strontium enters the system through rainfall and leaves the ecosystem via run-off into streams and deep drainage. In a study of strontium cycling in a Puerto Rican rain forest [4], the amount of strontium in the soil compartment was $\hat{x}_3 = 8.86$ kg/hectare. This number was computed by estimating the average soil depth (about 25 cm) and then the mass of the soil (1.24×10^6 kg/hectare) and by multiplying by the strontium concentration in the soil (7.15 p.p.m.). The run-off flow rate was estimated to be $F_{03} = 2.58 \times 10^5$ kg water/hectare/week and, from soil water samples beneath the root zone and stream samples after rain-storms, the strontium concentration was estimated to be $\hat{c}_3 = 0.01227$ p.p.m. (Fig. 11.1).

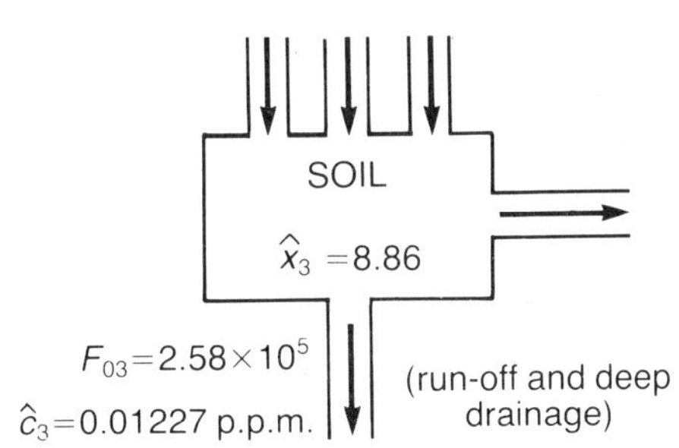

Fig. 11.1: Flows in and out of the soil compartment

The flux r_{03} is then estimated to be $F_{03} \hat{c}_3 = 0.00316$ kg/hectare/week and, using our estimate for $\hat{x}_3$, we obtain $a_{03} = 0.00316/8.86 = 0.000357$/week.

Example 11.2 In a study of the interactions of various fish and invertebrates on George's Bank in the Northwest Atlantic [8], G. G. Walter modeled this marine ecosystem with nine compartments, two of which were herring and cod. For the particular species of cod in the area, herring constitutes a substantial portion of the diet. The transfer coefficient a_{21} was estimated from the *gut contents* of samples of cod. An examination of the stomach contents revealed that about 16% of the diet is herring and that a cod consumes in the course of a year about 3.2 times its weight. (Not all of the digested herring can be converted into growth. The growth efficiency rate is estimated to be 19%.)

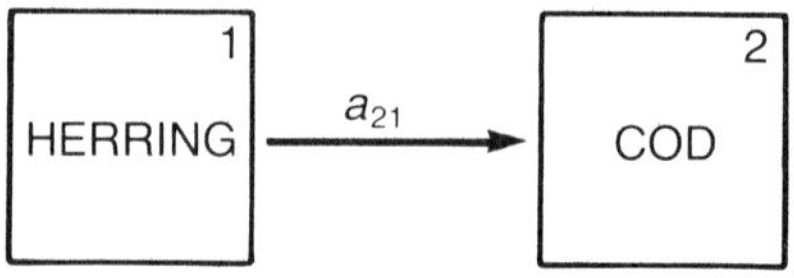

Fig. 11.2: Estimating a transfer coefficient from stomach contents

It follows that $r_{21} = (0.16)(3.2)\hat{x}_2 = 0.512\,\hat{x}_2$. Estimates for the biomass of cod and herring are $\hat{x}_2 = 2 \times 10^8$ kg and $\hat{x}_1 = 5 \times 10^8$ kg respectively, and so

$$a_{21} = \frac{r_{21}}{\hat{x}_1} = \frac{0.512\,\hat{x}_2}{\hat{x}_1} \approx 0.2/\text{year}$$

The next two examples illustrate indirect methods for estimating turnover rates.

Example 11.3 Following a nuclear cratering experiment carried out in July 1962 at the Nevada test site, ^{89}Sr concentrations on fallout-contaminated plants and ^{89}Sr concentrations in the bone ash of rabbits in the area were collected over a 60-day period. Martin and Turner [5] used this data and the two-compartment model shown in Fig. 11.3 to make predictions about the eventual half-lifes in the system. Assuming that the blast quickly deposits the strontium on the vegetation, then the initial condition takes the form $\mathbf{X}(0) = [x_0, 0]$. The model differential equations were solved in Example 8.7 where it was shown that

$$x_1(t) = x_0\, e^{-\lambda_1 t}$$

$$x_2(t) = \frac{\lambda_1 x_0}{\lambda_2 - \lambda_1}\,[e^{-\lambda_1 t} - e^{-\lambda_2 t}]$$

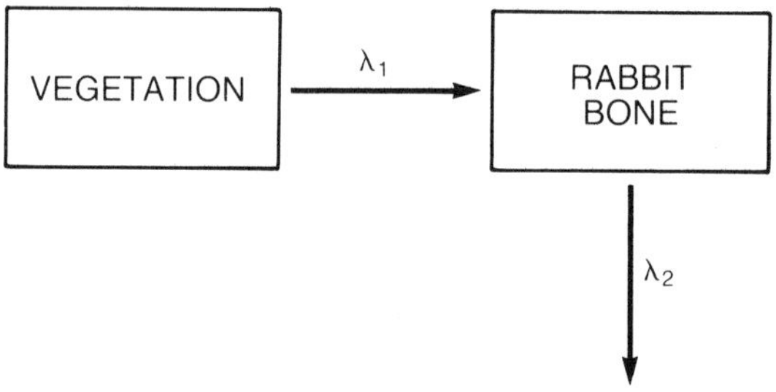

Fig. 11.3: Strontium cycling after fallout

Here x_0 is the amount of ^{89}Sr that will eventually be deposited in rabbit bone. From the data shown in Fig. 11.4, λ_1 is estimated to be 0.0385 and so the half-life in the vegetation is ln 2/0.0385 = 18 days.

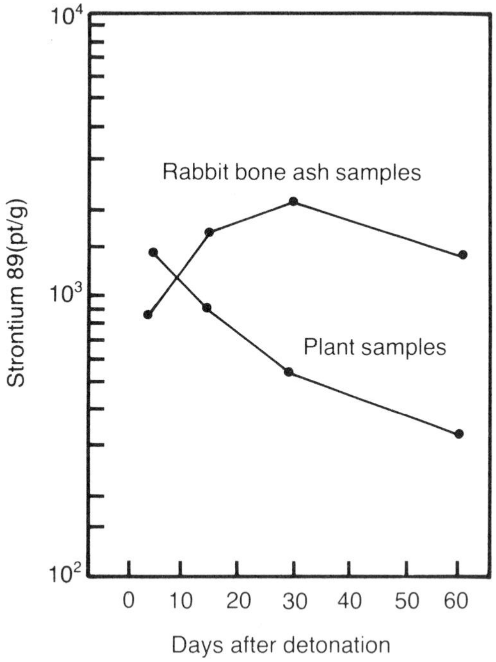

Fig. 11.4: Strontium concentrations in plant and bone samples

There are not enough data points, however, to perform the method of exponential peeling on the bone ash concentrations. We do know that the maximum concentration occurs at time $t = (\lambda_2 - \lambda_1)^{-1} \ln(\lambda_2/\lambda_1)$, and from Fig. 11.4, this occurs after about 30 days. We can estimate λ_2 by solving the equation

$$\frac{1}{\lambda_2 - 0.0385} \ln(\lambda_2/0.0385) = 30$$

for λ_2. Using Newton's method, for example, we obtain $\lambda_2 \approx 0.03$.

When the system has reached steady state $\hat{\mathbf{X}}$, then

$$0 = \dot{x}_i = \sum_{j \neq i} r_{ij}(\hat{\mathbf{X}}) - \sum_{j \neq i} r_{ji}(\hat{\mathbf{X}}) \quad \text{for each } i$$

Hence $\sum_{j \neq i} r_{ij} = \sum_{j \neq i} r_{ji}$, that is, the sum of the fluxes into a compartment equals the sum of the fluxes out of the compartment. When all but one of the fluxes is known, this unknown flux can be computed from the equation. For example, in Fig. 11.5, if r_{i1}, r_{i2}, and r_{3i} are known, then $r_{4i} = (r_{i1} + r_{i2}) - r_{3i}$, and then $a_{4i} = r_{4i}/\hat{x}_i$. This indirect method of computing a transfer coefficient is illustrated in our next example.

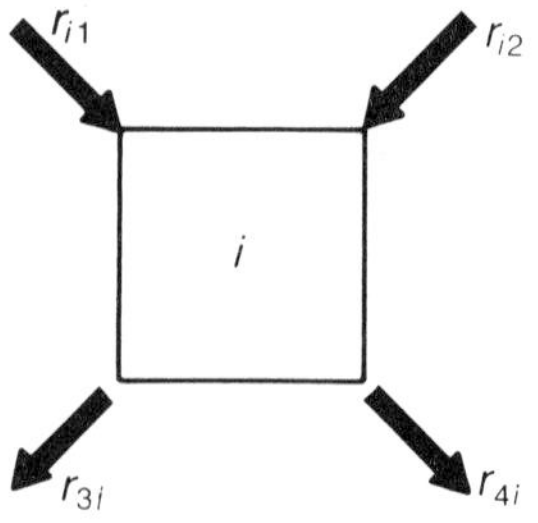

Fig. 11.5: At steady state, total flux in = total flux out.

Example 11.4 In studying strontium cycling in a forest ecosystem, it is difficult to measure directly the flux of strontium from the soil through the roots into the wood. If we are willing to assume steady state conditions, then the flux rate can be computed from others entering and leaving the soil as depicted in Fig. 11.6.

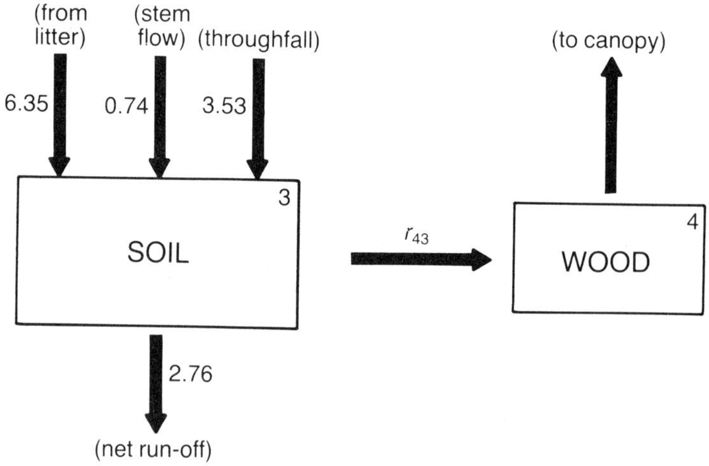

Fig. 11.6: Estimating an unknown flux rate (in g/hectare/week)

Thus $r_{43} + 2.76 = 6.35 + 0.74 + 3.53 = 10.62$ or $r_{43} = 7.86$ g/hectare/week.

11.3 STRONTIUM DYNAMICS IN A TROPICAL RAIN FOREST

A compartmental model for the cycling of radioactive ^{90}Sr in a rain forest is shown in Fig. 11.7. Except for the 0.002/month transfer coefficients, which result from losses due to natural radioactive decay, all other transfer coefficients were computed by measuring fluxes of stable strontium under assumed steady state conditions. As described by Jordan *et al.* [4], the top canopy of the forest extends from 20 to 25 meters off the ground. Except for the transfer from the

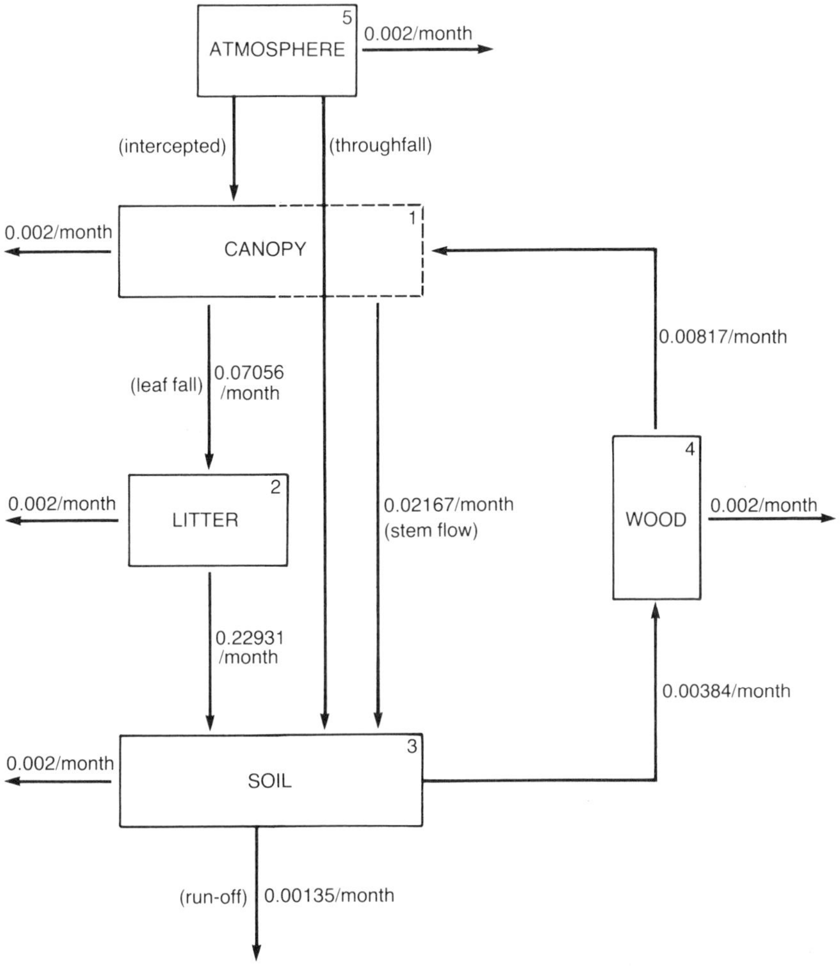

Fig. 11.7: Compartment diagram for strontium cycling in a tropical rain forest

canopy to litter by leaf, twig, and fruit-fall, water is the only important mechanism for the movement of strontium. When deposited into the forest ecosystem by rainfall, part of the strontium is intercepted and absorbed by the canopy while the remainder either falls directly to the soil ('throughfall') or runs slowly down the trees ('stem flow'). The following quotes from Jordan *et al.* describe some of the elaborate experiments necessary to estimate fluxes and compartment masses.

> For concentration determinations, leaf, stem, and root samples were collected in July and December 1968 from the six most common tree species of the forest at six different sites. At each site, four wood samples, four leaf samples, and one root sample were collected from each species. Litter and

soil samples were collected at each of the six sites in July only. At each site, four litter (including decaying organic matter) samples, and eight soil samples, four from 0 to 6 cm depth, and four from a 12 to 18 cm depth were taken. . .

Odum presents biomass data for the study area on a basis of weight per unit area of forest floor, for leaves, branches, stems and roots. These data are calculated from: regression equations developed from data obtained by cutting, drying, and weighing a series of trees from the area; a complete survey of all trees 4 inches and greater, diameter breast height, in each 2-in. size class, for a 8104 m^2 area. . .

Mass of organic matter on the forest floor was measured by collecting all material down to the mineral soil in 300 one-m^2 quadrats laid out randomly throughout the study area, and then oven-drying and weighing the samples. . .

Rainfall was measured continually with a standard Weather Bureau 'tipping-bucket' rain gauge mounted on top of a tower at 3.6 m above the canopy, and rainfall was recorded at the nearby field station. To measure throughfall, 15 trough-type rain gauges 1.5 m long by 5 cm wide by 30 cm deep were placed at various locations throughout the area. Depth of the water in the troughs was measured once a week for 2 yr. To measure stem flow, 30 trees of various sizes and species were fitted with polyvinyl tubing, scaled to the trees with paraffin. The tubing led down to collection barrels. The collected water was measured after every storm for 2 yr with a scaled dip stick. . .

Water samples for chemical analysis were taken weekly for 1 yr from the following collectors: (1) Two plastic barrels mounted on a tower 3.6 m above the canopy to collect rain samples; (2) Fifteen plastic barrels on the forest floor for throughfall collections (these barrels were covered with plastic screen to keep out leaves and litter); (3) the 30 stem flow collectors; (4) Eighteen soil water collectors located at the litter-soil interface, 13 at a 12.5 cm soil depth, and 6 at a 25 cm soil depth; (5) two samples collected from a nearby stream at low water; and (6) Two samples from the stream during high water, collected by anchoring plastic jars several feet above the normal low-water level so that they were filled during high water. . .

As a first application of the compartment model, suppose that rainfall quickly deposits a small quantity of ^{90}Sr directly into the soil and canopy compartments. This might occur when a thermonuclear device is used to excavate a canal or harbor near the forest. If we assume that the canopy intercepts 16% of the rainfall, then we must solve $\dot{\mathbf{X}} = \mathbf{A}\,\mathbf{X}$ where

$$\mathbf{A} = \begin{bmatrix} -0.09423 & 0 & 0 & 0.00817 \\ 0.07056 & -0.23131 & 0 & 0 \\ 0.02167 & 0.22931 & -0.00719 & 0 \\ 0 & 0 & 0.00384 & -0.01017 \end{bmatrix}$$

and $\mathbf{X}(0) = x_0\,[0.16, 0, 0.84, 0]$.

The solution is of the form

$$\begin{aligned} \mathbf{X}(t)/x_0 &= 0.0827\ \mathbf{E}_1\ e^{-0.2314t} + 0.3211\ \mathbf{E}_2\ e^{-0.0029t} \\ &\quad + 0.0646\ \mathbf{E}_3\ e^{-0.0150t} + 0.1625\ \mathbf{E}_4\ e^{-0.0936t} \end{aligned}$$

where eigenvectors $\mathbf{E}_1$, $\mathbf{E}_2$, $\mathbf{E}_3$, and $\mathbf{E}_4$ are columns of the matrix

$$\mathbf{Z} = \begin{bmatrix} 0.0011 & 0.0885 & -0.5243 & 1.0179 \\ -0.9978 & 0.0273 & -0.1711 & 0.5217 \\ 1.0205 & 1.8858 & 6.4448 & -1.6392 \\ -0.0177 & 0.9894 & -5.0822 & 0.0754 \end{bmatrix}$$

and the predicted levels of ^{90}Sr in the four compartments are graphed in Fig. 11.8.

A maximum of 0.8515 is reached in the soil after one year, while the wood compartment reaches a maximum of 0.1731 in the eleventh year. Notice that the radioactivity levels in the canopy and litter compartments decrease rapidly but then reach secondary relative maxima after twelve years. For large values of t, $\mathbf{X}(t)/x_0 \approx 0.3211\ \mathbf{E}_2\ e^{-0.0029t}$. It follows that the eventual half-life in all compartments is about twenty years.

To model radioactive fallout from the atmosphere following a large-scale nuclear test, we must add the atmosphere as a fifth compartment and use fallout data to estimate a_{15} and a_{35}. The model predicts a simple exponential decay from the atmosphere following the initial deposit of ^{90}Sr. Shown in Table 11.1 are average fallout rates for the years 1962–8.

Table 11.1

Year	Average fallout rate in San Juan (nanocuries/m^2/month)	Average fallout rate in the forest (nanocuries/m^2/month)
1962–3	0.534	0.88
1963–4	0.5988	0.99
1964–5	0.5111	0.84
1965–6	0.3771	0.62
1966–7	0.1750	0.29
1967–8	0.0466	0.08

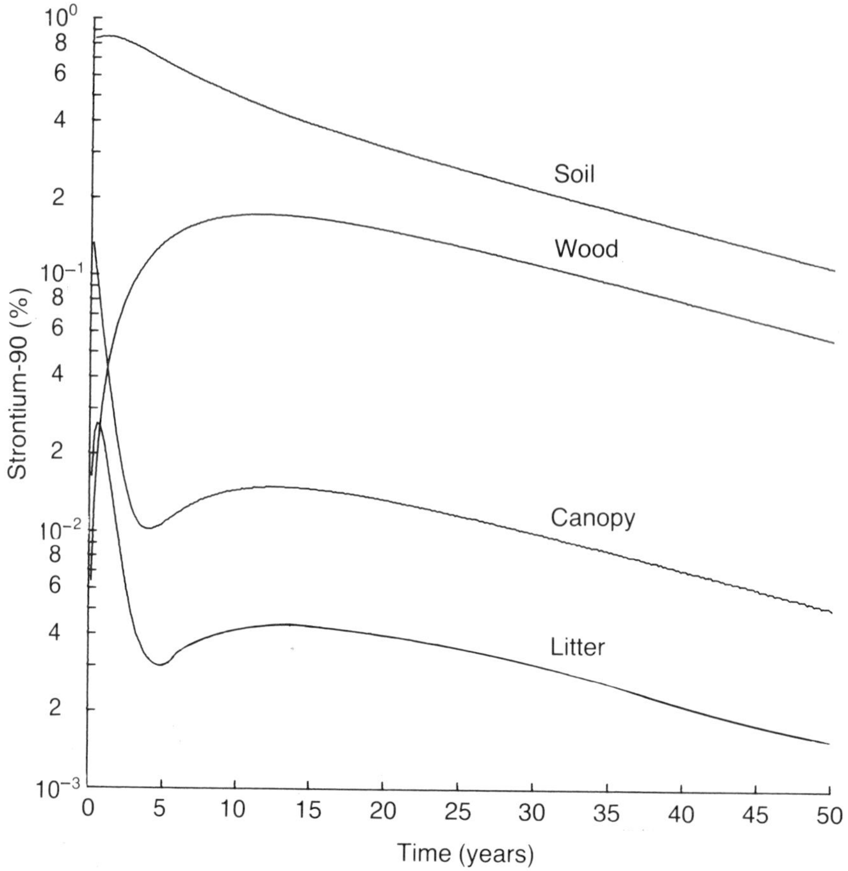

Fig. 11.8: Instantaneous injection into canopy and soil

Fitting an exponential function $\alpha\, e^{-\beta t}$ to this data gives $\beta = 0.05079$/month. Again, assuming that 16% of the rainfall is intercepted by the canopy, we set $a_{15} = (0.16)\,(0.05079) = 0.00813$ and $a_{35} = (0.84)\,(0.05079) = 0.04267$. With the initial condition now of the form $\mathbf{X}(0) = x_0\,[0, 0, 0, 0, 1]$, the solution is

$$\begin{aligned}\mathbf{X}(t)/x_0 \;=\; & -0.0235\,\mathbf{E}_1\, e^{-0.2314t} + 0.3265\,\mathbf{E}_2\, e^{-0.0029t} \\ & + 0.0869\,\mathbf{E}_3\, e^{-0.0150t} - 0.2022\,\mathbf{E}_4\, e^{-0.0936t} \\ & + \mathbf{E}_5\, e^{-0.0508t}\end{aligned}$$

where $\mathbf{E}_1$ through $\mathbf{E}_5$ are columns of the matrix

$$\mathbf{Z} = \begin{bmatrix} 0.0011 & 0.0885 & -0.5243 & 1.0179 & 0.2226 \\ -0.9978 & 0.0273 & -0.1711 & 0.5217 & 0.0880 \\ 1.0205 & 1.8858 & 6.4448 & -1.6392 & -1.4836 \\ -0.0177 & 0.9894 & -5.0822 & 0.0754 & 0.1336 \\ 0 & 0 & 0 & 0 & 1 \end{bmatrix}$$

and the solutions are graphed in Fig. 11.9.

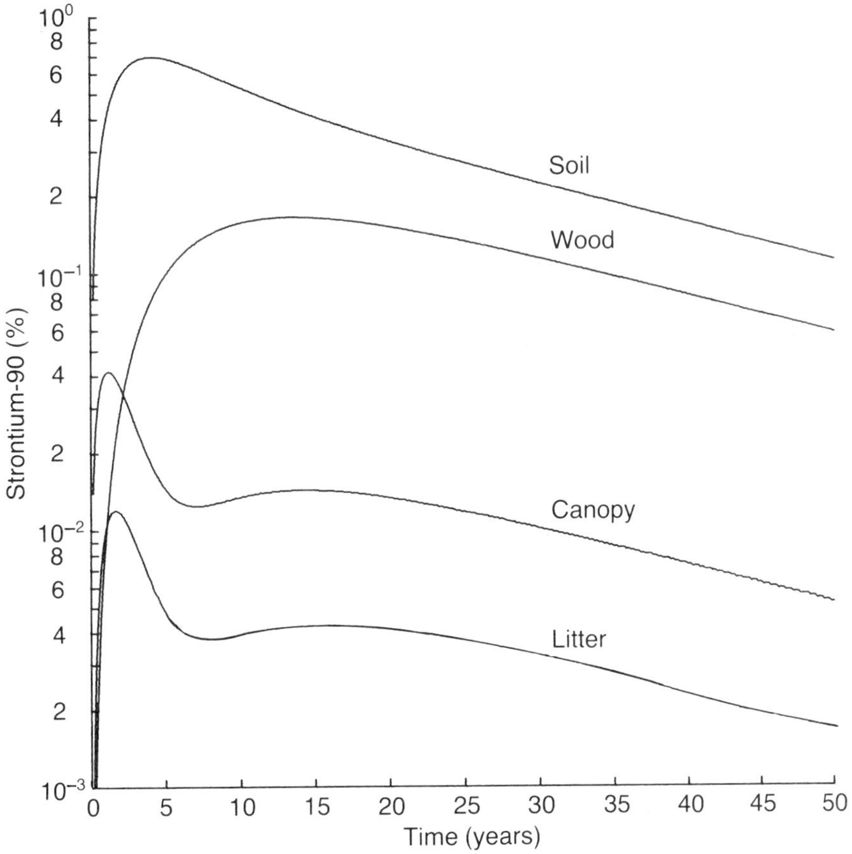

Fig. 11.9: Fallout from the atmosphere at the rate of 5% per month

A maximum of 0.695 is reached in the soil after four years, while the wood compartment shows a maximum of 0.165 during the thirteenth year. A maximum of 0.0416 in the canopy occurs during the second year and a secondary relative maximum of 0.0143 appears during the fourteenth year. For the litter compartment, the maximum of 0.0120 is also reached during the second year while a secondary maximum of 0.0044 occurs in year 15. For large t, $\mathbf{X}(t) \approx 0.3265\ \mathbf{E}_2\ e^{-0.0029t}$ and so the eventual half-life in the system is again about twenty years.

11.4 CARBON CYCLING IN THE ALEUTIAN ECOSYSTEM

The Aleutian Islands group consists of a chain of 70 islands extending west from the southern coast of Alaska a longitudinal distance of about 1200 miles. The entire ecosystem is extremely stable and productive. The nutrient-rich Pacific Ocean waters move into the Bering Sea and through the Aleutian chain making the system attractive to a large variety of sea mammals, fish, and birds. The native population, the Aleuts, have occupied this territory for over 8500 years. When discovered by Bering in 1741, the population numbered about 16 000 and this is the present day population figure. For centuries the Aleuts were highly dependent on the ecosystem for their food, shelter, and clothing, and their culture can only be understood when viewed from this perspective.

The following linear compartment model has been proposed as a *preliminary model* whose goal is to summarize the little quantitative information known about the ecosystem and to indicate areas where further research is needed. As an essential and stable component of all living things, carbon can be used as the tracer element. Shown in Fig. 11.10 is the compartmental diagram. The arrows between compartments represent fluxes of carbon (in metric tons/year). Except for the fluxes connected with man, these fluxes, together with compartmental masses, were obtained from rough world figures given by Bolin [1]. These world figures were multiplied by 0.025%, the percentage of the world's surface area occupied by the Aleutian system. It is possible that some of the figures are considerably different in the Aleutian system, but there is simply no data available. To estimate the amount of carbon in the Aleut population, the biomass of the population was estimated by summing average weights over the age classes. This figure was then multiplied by 17.5%, the percentage of carbon in the human body. Fluxes of carbon from the sea and from land were estimated using average daily food intakes and the fact that 95% of the diet comes from the sea. The flux $r_{34} = 46.7$ was computed from excretory data while the final flux r_{14} (respiration) was computed from the equation $r_{14} = (r_{42} + r_{46}) - r_{34}$. Assuming steady state has been reached, we may form the transfer matrix $\mathbf{A}$ shown below.

Letting $\mathbf{X}_1 = [1.75 \times 10^8, 1.12 \times 10^8, \ldots, 8.62 \times 10^9]$, then $\mathbf{A}\,\mathbf{X}_1 = 0$ since $\sum_{j \neq i} r_{ij} = \sum_{j \neq i} r_{ji}$ for each i. Since $\mathbf{A}$ is an ecomatrix, $\lambda_1 = 0$ is an eigenvalue of $\mathbf{A}$. The other eight eigenvalues are all negative with $\lambda_2 = -0.0067$ being the largest. Since $\mathbf{X}_1$ is an eigenvector corresponding to $\lambda_1 = 0$, it follows that the solutions to $\dot{\mathbf{X}} = \mathbf{A}\,\mathbf{X}$ are of the form

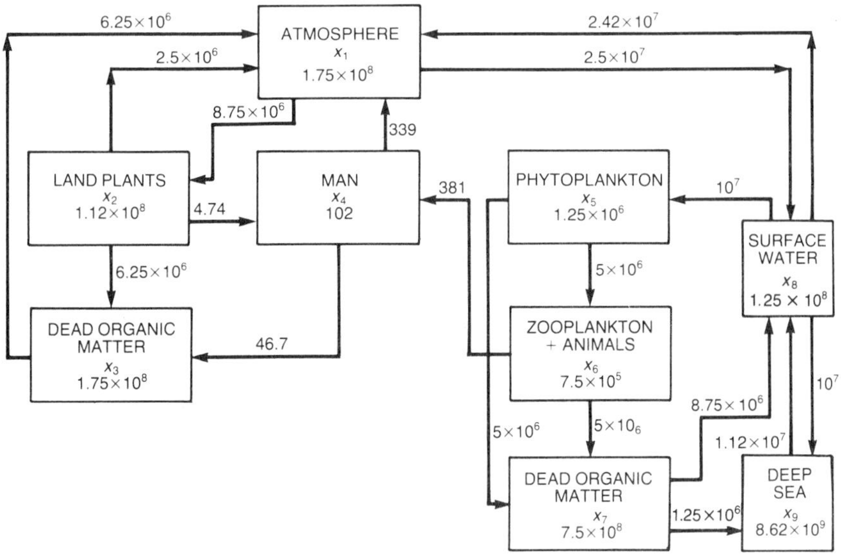

Fig. 11.10: Compartment diagram for carbon cycling in the Aleutian ecosystem

$$\begin{bmatrix}
-0.193 & 0.0223 & 0.0357 & 3.32 & 0 & 0 & 0 & 0.194 & 0 \\
0.05 & -0.0781 & 0 & 0 & 0 & 0 & 0 & 0 & 0 \\
0 & 0.0558 & -0.0357 & 0.458 & 0 & 0 & 0 & 0 & 0 \\
0 & 4.21\times10^{-8} & 0 & -3.78 & 0 & 5.08\times10^{-4} & 0 & 0 & 0 \\
0 & 0 & 0 & 0 & -8 & 0 & 0 & 0.08 & 0 \\
0 & 0 & 0 & 0 & 4 & -6.67 & 0 & 0 & 0 \\
0 & 0 & 0 & 0 & 4 & 6.67 & -0.013 & 0 & 0 \\
0.143 & 0 & 0 & 0 & 0 & 0 & 0.0113 & -0.354 & 0.0013 \\
0 & 0 & 0 & 0 & 0 & 0 & 0.0167 & 0.08 & -0.0013
\end{bmatrix}$$

Fig. 11.11: The transfer matrix **A**.

$$\mathbf{X}(t) = c_1\,\mathbf{X}_1 + c_2\mathbf{E}_2\,e^{-0.0067t} + \ldots + c_9\mathbf{E}_9\,e^{-\lambda_9 t}$$

and so, for large t, $\mathbf{X}(t) \approx c_1\mathbf{X}_1 + c_2\mathbf{E}_2\,e^{-0.0067t}$. This expression gives us some indication of how quickly the ecosystem recovers from a perturbation from $\mathbf{X}_1$ to $\mathbf{X}(0) = \mathbf{X}_0$, and also points out the fact that the *new equilibrium state need not be* $\mathbf{X}_1$. (Note that $e^{-0.0067t} = \frac{1}{2}$ for $t \approx 103$ years.) If, for example, the mass of phytoplankton is reduced by a factor of 10, then the new equilibrium state is $c_1\mathbf{X}_1$, where $c_1 = 0.99989$ and it takes only about a year for the phytoplankton mass to return to a value near 1.25×10^6.

Using this model, we can inquire into how sensitive the Aleut population is to minor changes in the other components of the system and to minor changes in various transfer coefficients. Although we could solve the system $\dot{\mathbf{X}} = \mathbf{A}\,\mathbf{X}$ again and again subject to a variety of changes in matrix entries and initial conditions, there is a systematic way of answering such questions based on *sensitivity analysis* formulae. To perform a sensitivity analysis on the model, we must compute *partial derivatives* of the form $\partial \mathbf{X}/\partial a_{ij}$, $\partial \mathbf{X}/\partial x_{i0}$, and $\partial \lambda/\partial a_{ij}$. We will return in Chapter 12 to this Aleutian ecosystem model after we have developed methods for finding these partial derivatives.

The Aleutian ecosystem model presented is extremely simplistic and many model assumptions including the donor-controlled hypothesis involving man can be questioned. We must remember, however, that modeling is an iterative process and we have presented only step one in the iteration. Once more information has been gathered about this system, model assumptions can be altered and more realistic models developed.

EXERCISES

Part A

In each of the following three problems, compute the transfer coefficients from the given information. State all assumptions that must be made in order to perform the computation.

1 In a model for energy flow in the marine community of the English channel [2], the community is divided into six main components: phytoplankton, zooplankton, pelagic fish, benthic fauna, demersal fish, and bacteria. Energy fluxes (in kilocalories/m^2/year) into and out of the pelagic fish compartment are shown below:

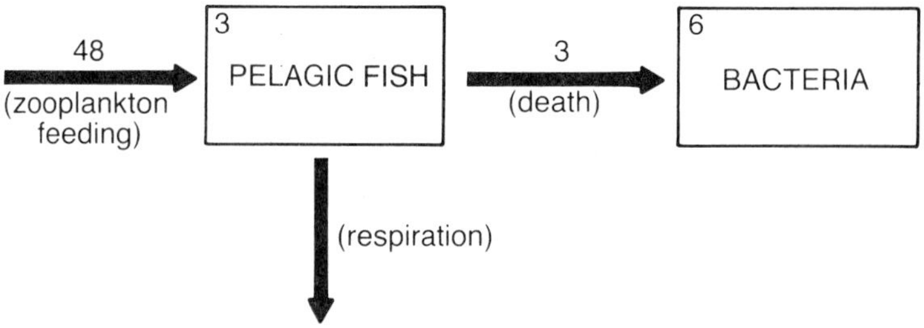

The annual standing crop of pelagic fish is 9.9 kilocalories/m^2.

(a) Find the transfer coefficient a_{63}.
(b) Find the respiration loss (in kcal/m^2/year) and then a_{03}.

2 The following information on the loss of DDT from the soil of agricultural land in the US is provided by Woodwell *et al.* [11]. Agricultural land retained 1.42×10^{11} grams of DDT when the rate of use of DDT was 2.7×10^{10} grams/year. Estimate the transfer coefficient out of the soil compartment.

3 In a compartment model for the production of needlerush (*Juncus roemerianus*) in a North Carolina salt-marsh [10], the leaves are classified as

living, dying, and dead, and the simple three-compartment model shown below was used:

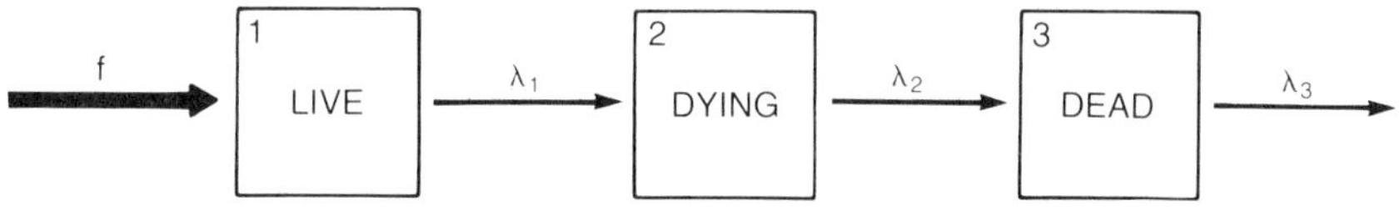

The average standing crops of live, dying, and dead leaves per m^2 were 344, 504, and 1604 grams respectively, and the annual growth rate was 792 grams/ m^2/year. Using this information, estimate λ_1, λ_2, and λ_3.

4 In a study of ^{137}Cs transfer in a poplar tree forest [7], 467 mCi of ^{137}Cs, placed in a trough encircling the tree trunks, enters the trees through chisel cuts. The compartmental diagram and transfer coefficients are shown in the diagram below. The unit is $(\text{day})^{-1}$.

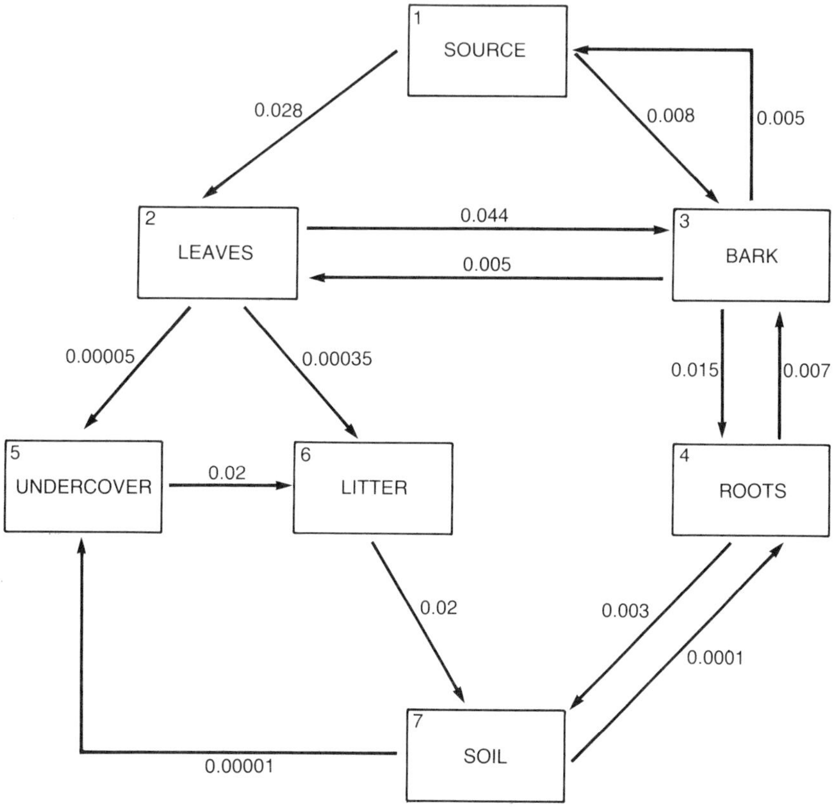

(a) Predict the maximum ^{137}Cs levels in the bark, leaves, and roots.
(b) Show that the amount of ^{137}Cs in the soil increases.
(c) Find $\hat{\mathbf{X}} = \lim_{t\to\infty} \mathbf{X}(t)$.

5 In a study of radiophosphorus cycling in a small plankton community [9], 100 μCi of ^{32}P were dissolved in a 200 liter aquarium. The eight compartments together with the transfer coefficients (in hr^{-1}) are shown below.

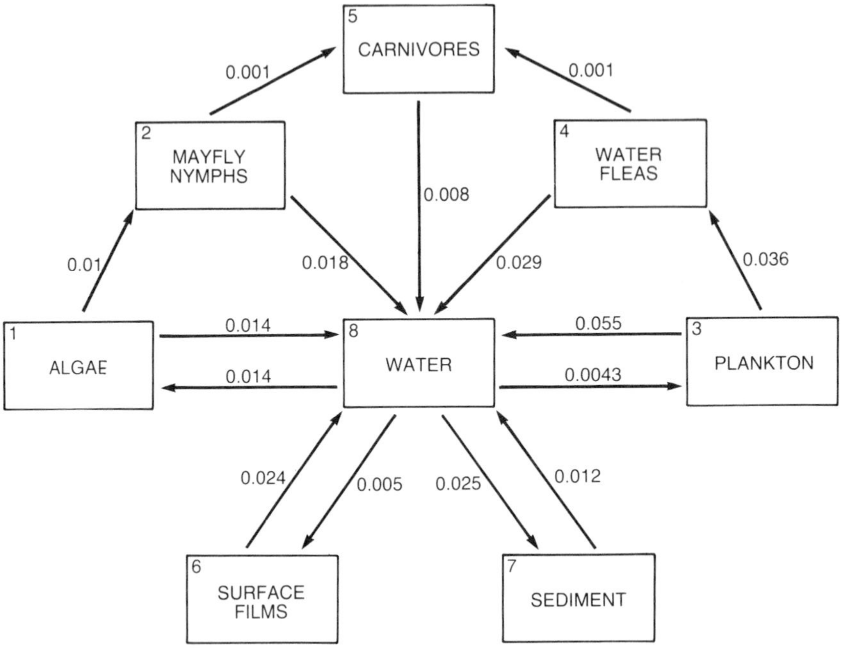

(a) How quickly does the ^{32}P move into the plants? Graph $x_1(t)$, $x_3(t)$, and $x_8(t)$.
(b) Describe $\mathbf{X}(t)$ for large t.

6 In the Aleut ecosystem model, find the new equilibrium state if the biomass of land plants were suddenly reduced by a factor of 10. How quickly would this new state be 'reached' (to 3 significant digits)?

REFERENCES

[1] B. Bolin, 'The Carbon Cycle', *Scientific American* (1970), **223**, 125–132.
[2] M. Brylinsky, 'Steady-State Sensitivity Analysis of Energy Flow in a Marine Ecosystem', in *Systems Analysis and Simulation in Ecology*, Volume II, p. 81–101, Academic Press, 1972.
[3] J. M. Hett and R. V. O'Neill, 'Systems Analysis of the Aleut Ecosystem', *Arctic Anthropology* (1974), Vol. XI (No. 1), 31–40.

[4] Carl F. Jordan, Jerry R. Kline, and Donald S. Sasscer, 'A Simple Model of Strontium and Manganese Dynamics in a Tropical Rain Forest', *Health Physics* (1973), **24**, 477–489.

[5] W. E. Martin and F. B. Turner, 'Transfer of ^{89}Sr from Plants to Rabbits in a Fallout Field', *Health Physics* (1966), **12**, 621–631.

[6] R. V. O'Neill, 'A Review of Linear Compartmental Analysis in Ecosystem Science', in *Compartmental Analysis of Ecosystem Models*, International Co-operative Publishing House, Fairland, Maryland, 1979, p. 3–28.

[7] Jerry S. Olsen, 'Equations for Cesium Transfer in a Liriodendron Forest', *Health Physics* (1965), **11**, 1385–1392.

[8] G. G. Walters, 'A Compartmental Model of a Marine Ecosystem', in *Compartmental Analysis of Ecosystem Models*, International Co-operative Publishing House, Fairland, Maryland, 1979, p. 29–42.

[9] R. H. Whittaker, 'Experiments with Radiophosphorus Tracer in Aquarium Microcosms', *Ecological Monographs* (1961), **31**(2), 157–188.

[10] Richard B. Williams and Marianne B. Murdoch, 'Compartmental Analysis of the Production of Juncus Roemerianus in a North Carolina Salt Marsh', *Cheasapeake Science* (1972), **13**(2), 69–79.

[11] George M. Woodwell, Paul P. Craig, and Horton A. Johnson, 'DDT in the Biosphere: Where Does It Go', *Science* (1971), **174**, 1101–1107.

FURTHER READING

R. E. Funderlic and M. T. Heath, *Linear Compartment Analysis of Ecosystems*, Oak Ridge National Laboratory (Oak Ridge, Tenn.), 1971.

Bernard C. Patten (Ed.), *Systems Analysis and Simulation in Ecology*, Volumes I–V, Academic Press (N.Y.).

R. W. Poole, *An Introduction to Quantitative Ecology*, McGraw-Hill (N.Y.), 1974.

R. H. Whittaker, *Communities and Ecosystems*, Macmillan Publishing Co. (N.Y.), 1975.

CHAPTER 12

Sensitivity Analysis

12.1 INTRODUCTION

As we have seen in Chapter 11, the key parameters in a linear ecosystem model must be estimated before we can make useful predictions about the future states in the system. Often the model requires more information than is presently available to estimate fluxes and biomasses, and we must make educated guesses as to these values. This was demonstrated in the Aleutian ecosystem model of Chapter 11. In other cases we use experimental data to estimate parameters and so we must live with standard experimental error. In general, model parameters are subject to uncertainty and these inaccuracies can throw off our numerical predictions and the resulting conclusions we draw from the models. In the Leslie population model, for example, errors in the survival and fecundity rates will produce errors in estimating the dominant eigenvalue and stable age distribution. In Chapter 9, we used curve fitting techniques to obtain estimates for turnover rates and used these estimates to design drug–dose schemes. Errors in estimating these rates can result in incorrect predictions of drug concentration levels.

One method of assessing the effect of these errors is to run the model again and again subject to a variety of perturbations of the model parameters α_1, $\alpha_2, \ldots, \alpha_k$. This method is somewhat subjective since the k model parameters can vary over a region Λ in k-dimensional space and we must select a finite set of parameter points in Λ. A second method, based on the computation of key partial derivatives, is called *sensitivity analysis.*

Suppose that $\alpha_1, \alpha_2, \ldots, \alpha_k$ are the model parameters and $\mathbf{X} = \mathbf{X}(t, \alpha_1, \alpha_2, \ldots, \alpha_k)$ is the resulting state at time t. If the parameters change by *small amounts* $\Delta\alpha_1, \Delta\alpha_2, \ldots, \Delta\alpha_k$, then, by the total increment formula,

$$\Delta\mathbf{X} = \mathbf{X}(t, \alpha_1 + \Delta\alpha_1, \ldots, \alpha_k + \Delta\alpha_k) - \mathbf{X}(t, \alpha_1, \ldots, \alpha_k)$$

$$\approx \frac{\partial \mathbf{X}}{\partial \alpha_1}\Delta\alpha_1 + \frac{\partial \mathbf{X}}{\partial \alpha_2}\Delta\alpha_2 + \ldots + \frac{\partial \mathbf{X}}{\partial \alpha_k}\Delta\alpha_k$$

where the partials are evaluated at $(t, \alpha_1, \ldots, \alpha_k)$. Note that the larger the value of $|\partial\mathbf{X}/\partial\alpha_i|$, the more the term $(\partial\mathbf{X}/\partial\alpha_i)\,\Delta\alpha_i$ can contribute to the overall error in $\mathbf{X}$.

We call $\partial\mathbf{X}/\partial\alpha_i$ the *sensitivity of* $\mathbf{X}$ *to* α_i at time t. If $\partial\mathbf{X}/\partial\alpha_i$ is large, then the only way this contribution to the overall error can be made small is by reducing $\Delta\alpha_i$. If $\partial\mathbf{X}/\partial\alpha_i$ is small, we need not worry as much about $\Delta\alpha_i$. Hence performing a sensitivity analysis can indicate to the model builder which parameters need special care in their estimation. The definition of sensitivity is illustrated in the following two examples.

Example 12.1 In Example 8.1, we used a single compartment model to estimate pollution levels in lakes after further pollution had ceased. If $x(t)$ is the amount of pollution at time t, then

$$x(t) = x_0\, e^{-(F/V)t}$$

and so $\partial x/\partial F = -(t/V)\,x(t)$ and $\partial x/\partial V = (F/V^2)\,t\,x(t)$. If our goal is to predict $x(t)$ after ten years, then, using $V = 458$ km^3 and $F = 175$ km^3/yr, $\partial x/\partial F = -0.000478\,x_0$ and $\partial x/\partial V = 0.0001827\,x_0$. If the estimates for V and F are off by as much as 10%, then

$$|\Delta x| \approx \left|\frac{\partial x}{\partial F}\Delta F + \frac{\partial x}{\partial V}\Delta V\right| \leqslant 0.0167\,x_0$$

Note that in general the relative sensitivity $(\partial x/\partial F)/(\partial x/\partial V)$ is $-(V/F) = 2.617$.

Example 12.2 For the two-compartment model in Example 8.7, we showed that the maximum concentration in the tissues was $X = x_0(\lambda_1/\lambda_2)\,e^{-\lambda_1 T}$ where $T = (\lambda_2 - \lambda_1)^{-1}\ln(\lambda_2/\lambda_1)$ and x_0 is the initial drug dose. Now

$$\frac{\partial X}{\partial \lambda_1} = (x_0/\lambda_2)\,e^{-\lambda_1 T} + x_0(\lambda_1/\lambda_2)\,e^{-\lambda_1 T}$$

$$\left[\frac{1}{\lambda_2 - \lambda_1} - \frac{\lambda_2}{(\lambda_2 - \lambda_1)^2}\ln(\lambda_2/\lambda_1)\right]$$

and

$$\frac{\partial X}{\partial \lambda_2} = (-x_0 \lambda_1/\lambda_2^2)\, e^{-\lambda_1 T} + X \left[\frac{\lambda_1}{(\lambda_2 - \lambda_1)^2} \ln(\lambda_2/\lambda_2) - \frac{\lambda_1}{\lambda_2 (\lambda_2 - \lambda_1)} \right]$$

In terms of X and T, the two partials are

$$\frac{\partial X}{\partial \lambda_1} = X/\lambda_1 + \frac{X}{\lambda_2 - \lambda_1} (1 - \lambda_2 T)$$

and

$$\frac{\partial X}{\partial \lambda_2} = -X/\lambda_2 + \frac{X \lambda_1}{\lambda_2 - \lambda_1} \left(T - \frac{1}{\lambda_2} \right)$$

If our estimates for the parameters are $\lambda_1 = 0.10$ and $\lambda_2 = 0.04$, then $T = 15.27$ and $X = 0.5429\ x_0$. Hence $\partial X/\partial \lambda_1 = 3.514\ X = 1.90788\ x_0$ and $\partial X/\partial \lambda_2 = -8.7858\ X = -4.7697\ x_0$. It follows that X is 2.5 times as sensitive to small changes in λ_2 than is λ_1. Unfortunately λ_2 is the harder of the two parameters to estimate.

Examples 12.1 and 12.2 are not typical. More typically we will not have explicit formulas for $\mathbf{X}(t)$ in terms of t and the parameters $\alpha_1, \ldots, \alpha_k$. In the remainder of the chapter we will develop the differential or difference equations that will be satisfied by the sensitivity vector

$$\mathbf{S}_i = \frac{\partial \mathbf{X}}{\partial \alpha_i} = \left[\frac{\partial x_1}{\partial \alpha_i}, \ldots, \frac{\partial x_n}{\partial \alpha_i} \right]$$

and we will then use the computer to find $\mathbf{S}_i$.

12.2 STATE SENSITIVITIES TO TRANSFER COEFFICIENTS

If $\mathbf{X}(t)$ is the solution to the linear system $\dot{\mathbf{X}} = \mathbf{A}\,\mathbf{X} + \mathbf{F}(t)$ with $\mathbf{X}(0) = \mathbf{X}_0$, then we can find the sensitivity of $\mathbf{X}(t)$ to a particular transfer coefficient a_{ij} by finding

$$\mathbf{S}_{ij} = \frac{\partial \mathbf{X}}{\partial a_{ij}} = \left[\frac{\partial x_1}{\partial a_{ij}}, \ldots, \frac{\partial x_n}{\partial a_{ij}} \right]$$

Here $\mathbf{A}$ is a transfer matrix and so $a_{jj} = -\sum_{k \neq j} a_{kj}$. Since $\partial \mathbf{X}/\partial t = \mathbf{A}\,\mathbf{X} + \mathbf{F}(t)$, we have

$$\frac{\partial}{\partial a_{ij}} \left(\frac{\partial \mathbf{X}}{\partial t} \right) = \frac{\partial \mathbf{A}}{\partial a_{ij}} \mathbf{X} + \mathbf{A} \frac{\partial \mathbf{X}}{\partial a_{ij}}$$

Assuming the mixed second partials are continuous, we can interchange the order of differentiation to obtain

$$\frac{\partial}{\partial t}\,\mathbf{S}_{ij} = \mathbf{A}\,\mathbf{S}_{ij} + \frac{\partial \mathbf{A}}{\partial a_{ij}}\,\mathbf{X}(t)$$

Let $\mathbf{Z}_{ij}$ denote the $n \times n$ matrix with all zeros except for a 1 in the (i, j)th position. Then since $\partial a_{jj}/\partial a_{ij} = -1$ for $i \neq j$, we see that the sensitivity vector $\mathbf{S}_{ij}$ is a solution to the non-homogeneous linear system

$$\boxed{\dot{\mathbf{S}}_{ij} = \mathbf{A}\,\mathbf{S}_{ij} + (\mathbf{Z}_{ij} - \mathbf{Z}_{jj})\,\mathbf{X}(t)}$$

and since $\mathbf{X}(0) = \mathbf{X}_0$ regardless of the value of a_{ij}, we have $\mathbf{S}_{ij}(0) = \mathbf{0}$. (Note that $\mathbf{S}_{0j}$ satisfies $\dot{\mathbf{S}}_{0j} = \mathbf{A}\,\mathbf{S}_{0j} - \mathbf{Z}_{jj}\mathbf{X}(t)$.) Since the driving function will involve terms of the form $\mathbf{E}_k\, e^{\lambda_k t}$, where λ_k is an eigenvalue of $\mathbf{A}$, it is not possible to construct a particular solution using Theorem 7.3. Although there are methods to cover this case (see Exercises 21 and 22 in Chapter 7), we will apply a numerical method to generate $\mathbf{X}(t)$ and $\mathbf{S}_{ij}(t)$ simultaneously. Let $\mathbf{Y}(t) = [\mathbf{X}(t) \,\vdots\, \mathbf{S}_{ij}(t)]$ and let $\mathbf{B}$ be the matrix

$$\left[\begin{array}{c:c} \mathbf{A} & \mathbf{0} \\ \hdashline \dfrac{\partial \mathbf{A}}{\partial a_{ij}} & \mathbf{A} \end{array}\right]$$

Then $\dot{\mathbf{Y}} = \mathbf{B}\,\mathbf{Y} + \begin{bmatrix} F(t) \\ 0 \end{bmatrix}$ and $\mathbf{Y}(0) = [\mathbf{X}_0, 0]$. In our next example, we will use the 4th order Runge–Kutta method on this system to generate $\mathbf{S}_{ij}$.

Example 12.3 For the ^{90}Sr dynamics model on pages 168–174, we will compute the state sensitivity to the various transfer coefficients at the ten-year mark ($t = 120$ months). To find $\partial\mathbf{X}/\partial a_{21}$, for example, we must solve $\dot{\mathbf{Y}} = \mathbf{B}\,\mathbf{Y}$, where $\mathbf{Y}(0) = [0.16, 0, 0.84, 0, 0, 0, 0, 0]$ and $\mathbf{B}$ is the matrix

$$\left[\begin{array}{cccc:cccc}
-0.09423 & 0 & 0 & 0.00817 & 0 & 0 & 0 & 0 \\
0.07056 & -0.23131 & 0 & 0 & 0 & 0 & 0 & 0 \\
0.02167 & 0.22931 & -0.00719 & 0 & 0 & 0 & 0 & 0 \\
0 & 0 & 0.00384 & -0.01017 & 0 & 0 & 0 & 0 \\
\hdashline
-1 & 0 & 0 & 0 & -0.09423 & 0 & 0 & 0.00817 \\
1 & 0 & 0 & 0 & 0.07056 & -0.23131 & 0 & 0 \\
0 & 0 & 0 & 0 & 0.02167 & 0.22931 & -0.00719 & 0 \\
0 & 0 & 0 & 0 & 0 & 0 & 0.00384 & -0.01017
\end{array}\right]$$

Shown in Table 12.1 are the resulting state sensitivities after ten years. Note that the highest sensitivities are associated with a_{43} and a_{03}. If the transfer coefficients can be in error by as much as 10%, then the resulting error in **X**, $\Delta\mathbf{X}$, can be approximated by

$$\mathbf{E} = \frac{\partial \mathbf{X}}{\partial a_{21}} \Delta a_{21} + \frac{\partial \mathbf{X}}{\partial a_{31}} \Delta a_{31} + \frac{\partial \mathbf{X}}{\partial a_{32}} \Delta a_{32} + \frac{\partial \mathbf{X}}{\partial a_{43}} \Delta a_{43} + \frac{\partial \mathbf{X}}{\partial a_{14}} \Delta a_{14} + \frac{\partial \mathbf{X}}{\partial a_{03}} \Delta a_{03}$$

and the components of **E** have absolute value $\leqslant$ [0.0036, 0.0029, 0.0278, 0.0227]. Since $\mathbf{X}(10)$ = [0.0146, 0.0044, 0.4986, 0.1716], our estimate for the soil compartment ranges from about 0.47 to 0.53.

Table 12.1 State sensitivities at year 10

Transfer coefficient	Canopy	Litter	Soil	Wood
a_{21} (canopy to litter)	−0.1478	0.0180	0.0765	0.0321
a_{31} (canopy to soil)	−0.1465	−0.0441	0.1131	0.0465
a_{32} (litter to soil)	0.0004	−0.0187	0.0110	0.0044
a_{43} (soil to wood)	3.1450	0.9562	−36.42	36.44
a_{14} (wood to canopy)	1.1245	0.3461	6.4971	−8.3630
a_{03} (run-off from soil)	−0.8238	−0.2389	−55.3337	−10.6413

12.3 STATE SENSITIVITY TO INITIAL CONDITIONS

If the initial condition $\mathbf{X}(0) = [x_{10}, x_{20}, \ldots, x_{n0}]$ is estimated from experimental data, then this uncertainty is another source for error in predicting $\mathbf{X}(t)$. To measure the sensitivity of **X** to changes in the initial conditions, we compute the partial derivatives $\partial\mathbf{X}/\partial x_{i0}$. If $\mathbf{X}(t)$ is a solution to $\dot{\mathbf{X}} = \mathbf{A}\,\mathbf{X} + \mathbf{F}(t)$ satisfying $\mathbf{X}(0) = \mathbf{X}_0$, then, from Theorem 7.2,

$$\mathbf{X}(t) = \mathrm{e}^{\mathbf{A}t}\,\mathbf{X}(0) + \int_0^t \mathrm{e}^{\mathbf{A}(t-s)}\,\mathbf{F}(s)\,\mathrm{d}s$$

Hence $\partial\mathbf{X}/\partial x_{i0} = \mathrm{e}^{\mathbf{A}t}\,(\partial\mathbf{X}(0)/\partial x_{i0})$. Assuming that the components of $\mathbf{X}(0)$ are independent, we have $\partial\mathbf{X}/\partial x_{i0} = \mathrm{e}^{\mathbf{A}t}\,\mathbf{Z}_i$, where $\mathbf{Z}_i = [0, \ldots 0, 1, 0, \ldots, 0]$ and 1 occurs in the ith position. Note that $\mathrm{e}^{\mathbf{A}t}\,\mathbf{Z}_i$ is just the solution to $\dot{\mathbf{Y}} = \mathbf{A}\,\mathbf{Y}$ which satisfies $\mathbf{Y}(0) = \mathbf{Z}_i$.

Example 12.4 For the rain forest model in which rainfall quickly deposits ^{90}Sr into the canopy and soil compartments, the initial condition was of the form $\mathbf{X}(0) = [\alpha x_0, 0, (1-\alpha)x_0, 0]$, where α determines the percentage of rainfall intercepted by the canopy. We used $\alpha = 0.16$ in the model, but this percentage is subject to some variation. Therefore we wish to compute $\partial\mathbf{X}/\partial\alpha$. Preceding as above with $x_0 = 1$,

$$\frac{\partial \mathbf{X}}{\partial \alpha} = e^{\mathbf{A}t}\,[1, 0, -1, 0]$$

If we let $t = 10$ years ($= 120$ in the model), we obtain

$$\frac{\partial \mathbf{X}}{\partial \alpha} = [-0.\,0009, -0.0003, 0.0197, -0.0087]$$

This indicates that the prediction $\mathbf{X}(10)$ is insensitive to small errors in estimating α.

12.4 STEADY STATE SENSITIVITIES TO TRANSFER COEFFICIENTS

When steady state solutions $\hat{\mathbf{X}} = \lim_{t \to \infty} \mathbf{X}(t)$ exist, we can compute $\partial \hat{\mathbf{X}} / \partial a_{ij}$ to measure the sensitivity of $\hat{\mathbf{X}}$ to small changes in a_{ij}. Since $\mathbf{S}_{ij} = \partial \mathbf{X} / \partial a_{ij}$ satisfies

$$\frac{\partial \mathbf{S}_{ij}}{\partial t} = \mathbf{A}\,\mathbf{S}_{ij} + \frac{\partial \mathbf{A}}{\partial a_{ij}}\,\mathbf{X}(t)$$

then the steady state sensitivity $\hat{\mathbf{S}}_{ij} = \partial \hat{\mathbf{X}} / \partial a_{ij}$ should satisfy $\mathbf{0} = \mathbf{A}\,\hat{\mathbf{S}}_{ij} + \partial \mathbf{A} / \partial a_{ij}\,\hat{\mathbf{X}}$ or

$$\boxed{\mathbf{A}\,\hat{\mathbf{S}}_{ij} = -\frac{\partial \mathbf{A}}{\partial a_{ij}}\,\hat{\mathbf{X}}}$$

Note that $\partial \mathbf{A} / \partial a_{ij} = \mathbf{Z}_{ij} - \mathbf{Z}_{jj}$ and $\partial \mathbf{A} / \partial a_{0j} = -\mathbf{Z}_{jj}$ and so

$$\mathbf{A}\,\hat{\mathbf{S}}_{ij} = \begin{cases} \hat{x}_j(\mathbf{Z}_j - \mathbf{Z}_i), & j \neq 0 \\ \hat{x}_j\,\mathbf{Z}_j, & j = 0 \end{cases}$$

If $\mathbf{A}$ is invertible, we have the unique solution $\hat{\mathbf{S}}_{ij} = -\mathbf{A}^{-1}\,\partial \mathbf{A} / \partial a_{ij}\,\hat{\mathbf{X}}$. Often, however, $\lambda = 0$ will be an eigenvalue of $\mathbf{A}$ and so $\mathbf{A}^{-1}$ will not exist. If this eigenvalue has multiplicity one, then solutions will be of the form $\hat{\mathbf{S}}_{ij} = \mathbf{S}_p + c_1\,\mathbf{E}_1$ where $\mathbf{S}_p$ is a particular solution, $\mathbf{E}_1$ is a non-zero eigenvector corresponding to $\lambda = 0$, and c_1 is an arbitrary constant. When the *system is closed*, no material is lost and so

$$\sum_{i=1}^{n} \hat{x}_i = \sum_{i=1}^{n} x_{io}$$

where $\mathbf{X}(0) = [x_{10}, x_{20}, \ldots, x_{no}]$. Hence

$$\boxed{\frac{\partial \hat{x}_1}{\partial a_{ij}} + \frac{\partial \hat{x}_2}{\partial a_{ij}} + \ldots + \frac{\partial \hat{x}_n}{\partial a_{ij}} = 0}$$

This additional equation can be used to determine c_1. Alternately we can use the last piece of information to reduce the system to $n - 1$ equations in $n - 1$ unknowns.

Example 12.5 In Example 9.2, we used estimates for turnover rates of the drug lidocaine to determine an infusion rate that would lead to a steady state concentration of 3.5 mg/liter in the bloodstream. If $\mathbf{X} = [x_1, x_2]$, then

$$\dot{\mathbf{X}} = \begin{bmatrix} -0.090 & 0.038 \\ 0.066 & -0.038 \end{bmatrix} \mathbf{X} + \begin{bmatrix} I \\ 0 \end{bmatrix}$$

where $I = 2.52$ mg/minute, and the steady state is $\hat{\mathbf{X}} = [105, 182.368]$. For the transfer coefficient a_{12},

$$\hat{\mathbf{S}}_{12} = \hat{x}_2 \, \mathbf{A}^{-1} \, (\mathbf{Z}_2 - \mathbf{Z}_1) = [0, -4799]$$

Likewise $\hat{\mathbf{S}}_{21} = [0, 2763]$ and $\hat{\mathbf{S}}_{01} = -[4375, 7599]$.

If the transfer coefficients can be in error by as much as 20%, then $\Delta\mathbf{X}$ can be approximated by

$$\mathbf{E} = \hat{\mathbf{S}}_{12} \, \Delta a_{12} + \hat{\mathbf{S}}_{21} \, \Delta a_{21} + \hat{\mathbf{S}}_{01} \, \Delta a_{01}$$

The components of $\mathbf{E}$ are no bigger than [21, 109.4] and so the steady state concentration in the bloodstream could be as large as (105 + 21)mg/30 liters = 4.2 mg/liter.

Example 12.6 For the Aleutian ecosystem model of Chapter 11, we can use the sensitivity formulas developed to see the effects of small perturbations of the transfer coefficients on those components affecting man. This system is closed and $\lambda = 0$ is an eigenvalue of multiplicity one. Hence, to solve the system $\mathbf{A}\,\hat{\mathbf{S}}_{ij} = -(\partial \mathbf{A}/\partial a_{ij})\,\hat{\mathbf{X}}$ for $\mathbf{S}_{ij}$, we will use the relationship $s_9 = -(s_1 + s_2 + \ldots + s_8)$, where $s_k = \partial \hat{x}_k / \partial a_{ij}$.

The first seven equations do not involve the variable s_9, and the last equation will be dropped. For equation (8), we eliminate s_9 using the relationship $0.0013\, s_9 = -0.0013(s_1 + \ldots + s_8)$. The new 8×8 matrix $\mathbf{B}$ for the system is now invertible with inverse (see page 188) and so $\mathbf{S} = -\mathbf{B}^{-1}\,\partial \mathbf{A}/\partial a_{ij}\,\hat{\mathbf{X}}$, where $\mathbf{S} = [s_1, \ldots, s_8]$.

For $\partial \hat{x}_4/\partial a_{42}$, we compute $\hat{\mathbf{S}}_{42} = \mathbf{B}^{-1}\,\hat{x}_2\,(\mathbf{Z}_2 - \mathbf{Z}_4)$ to obtain 2.96×10^7. From $\hat{\mathbf{S}}_{46} = \mathbf{B}^{-1}\,\hat{x}_6(\mathbf{Z}_6 - \mathbf{Z}_4)$, we find that $\partial \hat{x}_4/\partial a_{46} = 1.98 \times 10^5$. A 10% decrease in a_{42} will therefore result in $\Delta \hat{x}_4 \approx -0.125$ while a 10% decrease in a_{46} gives $\Delta \hat{x}_4 \approx -10.06$. This shows the extreme dependence of the Aleuts on the marine system.

The major carbon source for land plants is the atmosphere while surface water is the major source for marine life. To assess the effects in changes in the rate of uptake from these sources we compute $\partial \hat{x}_2/\partial a_{21}$ and $\partial \hat{x}_6/\partial a_{58}$.

$$
\mathbf{B}^{-1} = -\begin{bmatrix}
19.6714 & 19.1172 & 19.1984 & 19.5773 & 9.9899 & 9.9915 & 9.9925 & 12.9903 \\
12.5937 & 25.0430 & 12.2909 & 12.5335 & 6.3956 & 6.3966 & 6.3973 & 8.3164 \\
19.6844 & 39.1430 & 47.2223 & 22.9842 & 9.9967 & 9.9984 & 9.9992 & 12.9989 \\
7.67 \times 10^{-6} & 7.48 \times 10^{-6} & 7.39 \times 10^{-6} & 0.26456 & 1.61 \times 10^{-5} & 2.62 \times 10^{-5} & 6.01 \times 10^{-6} & 7.81 \times 10^{-6} \\
0.09345 & 0.08937 & 0.08997 & 0.0928 & 0.19863 & 0.07364 & 0.07365 & 0.09575 \\
0.05604 & 0.05359 & 0.05395 & 0.0556 & 0.11912 & 0.19409 & 0.04417 & 0.05742 \\
57.5095 & 54.9956 & 55.3638 & 57.0944 & 122.236 & 122.242 & 122.249 & 58.9241 \\
9.3453 & 8.9368 & 8.9966 & 9.2778 & 7.3634 & 7.3640 & 7.3655 & 9.5752
\end{bmatrix}
$$

Computing $\mathbf{B}^{-1}\,\hat{x}_1(\mathbf{Z}_1 - \mathbf{Z}_2)$ and $\mathbf{B}^{-1}\,\hat{x}_8(\mathbf{Z}_8 - \mathbf{Z}_5)$, we obtain $\partial\hat{x}_2/\partial a_{21} = 2.179 \times 10^9$ and $\partial\hat{x}_6/\partial a_{58} = 7.713 \times 10^6$. It follows that

$$
\frac{\partial\hat{x}_2/\partial a_{21}}{\partial\hat{x}_6/\partial a_{58}} \approx 280
$$

Thus the marine system, being far less sensitive to environmental changes, provides a much more reliable food source for the Aleuts than land plants.

12.5 EIGENVALUE SENSITIVITIES TO TRANSFER COEFFICIENTS

We next will develop formulas for $\partial\lambda_1/\partial a_{ij}$, where λ_1 is an eigenvalue of multiplicity one for a matrix **A**. In the physiological and ecological models we have presented, the transfer matrix **A** possessed a crucial eigenvalue λ_1 that determined the long range half-life of the tracer in the system. In addition, the dominant eigenvalue of a Leslie matrix **P** was an important numerical characteristic of a population. It is therefore important to assess the effects on λ_1 of small errors in estimating the matrix entries.

Let $D = \det(\lambda\mathbf{I} - \mathbf{A}) = (\lambda - \lambda_1)\,q(\lambda)$, where $q(\lambda_1) \neq 0$. Then, using the product rule

$$
\frac{\partial D}{\partial a_{ij}} = (\lambda - \lambda_1)\,\frac{\partial q(\lambda)}{\partial a_{ij}} - q(\lambda)\,\frac{\partial\lambda_1}{\partial a_{ij}}
$$

and

$$
\frac{\partial D}{\partial\lambda} = (\lambda - \lambda_1)\,\frac{\partial q}{\partial\lambda} + q(\lambda)
$$

Evaluating at $\lambda = \lambda_1$, we obtain $\partial D/\partial a_{ij} = -q(\lambda_1)\,\partial\lambda_1/\partial a_{ij}$ and $\partial D/\partial\lambda = q(\lambda_1)$. Since $q(\lambda_1) \neq 0$, we have

$$\boxed{\frac{\partial\lambda_1}{\partial a_{ij}} = -\frac{\partial D/\partial a_{ij}}{\partial D/\partial\lambda} \quad \text{at } \lambda = \lambda_1}$$

Both $\partial D/\partial a_{ij}$ and $\partial D/\partial\lambda$ can be computed in terms of the cofactors of the matrix $\lambda\mathbf{I} - \mathbf{A}$. Recall that the cofactor C_{ij} of a matrix $\mathbf{B}$ is defined to be $(-1)^{i+j}$ times the determinant of the matrix obtained by deleting the ith row and jth column from $\mathbf{B}$.

Expanding $\det(\lambda\mathbf{I} - \mathbf{A})$ down the jth column, we have

$$D = \sum_{k=1}^{n} (\lambda\delta_{kj} - a_{kj})\, C_{kj}$$

where δ_{kj} is the Kronecker delta. If $\mathbf{A}$ is a transfer matrix, none of the cofactors contain a_{ij} and so

$$\frac{\partial D}{\partial a_{ij}} = \sum_{k=1}^{n} -C_{kj}\,\frac{\partial a_{kj}}{\partial a_{ij}} = C_{jj} - C_{ij}$$

We next show that $\partial D/\partial\lambda = \sum_{k=1}^{n} C_{kk}$. To give the derivation, recall that if $\mathbf{B}(\lambda) = [b_{ij}(\lambda)]$, then

$$\mathbf{B}'(\lambda) = \frac{\partial\mathbf{B}}{\partial\lambda} = \mathbf{D}_1(\lambda) + \mathbf{D}_2(\lambda) + \ldots + \mathbf{D}_n(\lambda)$$

where $\mathbf{D}_k(\lambda)$ is obtained from $\mathbf{B}(\lambda)$ be replacing the kth row of $\mathbf{B}$ by its derivative $b'_{k1}(\lambda), \ldots, b'_{kn}(\lambda)$. On our case, $b_{kj}(\lambda) = \delta_{kj}\lambda - a_{kj}$ so that $b'_{kj}(\lambda) = \delta_{kj}$. Hence, the kth row of $\lambda\mathbf{I} - \mathbf{A}$ will be replaced by $\mathbf{Z}_k = [0, \ldots, 0, 1, 0, \ldots, 0]$, where 1 occurs in the kth position. Expanding $D_k(\lambda)$ along the kth row, we obtain the cofactor C_{kk} of $\lambda\mathbf{I} - \mathbf{A}$. Hence

$$\frac{\partial D}{\partial\lambda} = \sum_{k=1}^{n} C_{kk}$$

From our prior formula, we obtain for a *transfer matrix*

$$\boxed{\frac{\partial\lambda_1}{\partial a_{ij}} = \frac{C_{ij} - C_{jj}}{\sum_{k=1}^{n} C_{kk}}}$$

where C_{ij} is the (i, j)th cofactor of $\lambda_1 \mathbf{I} - \mathbf{A}$, and for transfer coefficients of the form a_{0j}, we have

$$\boxed{\frac{\partial \lambda_1}{\partial a_{0j}} = \frac{-C_{jj}}{\sum_{k=1}^{n} C_{kk}}}$$

(When $\mathbf{A}$ is not a transfer matrix and all entries are independent, then the formula is $\partial\lambda_1/\partial a_{ij} = -C_{ij} / \sum_{k=1}^{n} C_{kk}$.)

When λ_1 is an eigenvalue of multiplicity one, it can be shown that $C_{ij} = C_{i1}C_{1j}/C_{11}$ for $C_{11} \neq 0$. Thus all the cofactors can be computed from the $2n - 1$ cofactors $C_{11}, C_{12}, \ldots, C_{1n}, C_{21}, C_{31}, \ldots, C_{n1}$. These formulas are illustrated in our next two examples.

Example 12.7 Let $\mathbf{A}$ be the transfer matrix

$$\begin{bmatrix} -0.5 & 0 & 0 \\ 0.5 & -2 & 1 \\ 0 & 2 & -2 \end{bmatrix}$$

Then $\mathbf{A}$ has eigenvalues $\lambda = -1/2$, and $-2 \pm \sqrt{2}$. For $\lambda_1 = -1/2$,

$$\lambda_1 \mathbf{I} - \mathbf{A} = \begin{bmatrix} 0 & 0 & 0 \\ -0.5 & 1.5 & -1 \\ 0 & -2 & 1.5 \end{bmatrix}$$

and the cofactor matrix $\mathbf{C} = [C_{ij}]$ is easily seen to be

$$\mathbf{C} = \begin{bmatrix} 1/4 & 3/4 & 1 \\ 0 & 0 & 0 \\ 0 & 0 & 0 \end{bmatrix}$$

Hence $\partial\lambda_1/\partial a_{21} = (C_{21} - C_{11})/1/4 = -1$, while $\partial\lambda_1/\partial a_{23} = \partial\lambda_1/\partial a_{32} = \partial\lambda_1/\partial a_{03} = 0$. It follows that $\Delta\lambda_1 \approx -\Delta a_{21}$. (One can also verify directly that if $a_{21} = 0.5 + \epsilon$, then $\lambda_1 = 0.5 + \epsilon$.)

Example 12.8 The transfer matrix for the ^{90}Sr model in Chapter 11 was

$$\mathbf{A} = \begin{bmatrix} -0.09423 & 0 & 0 & 0.00817 \\ 0.07056 & -0.23131 & 0 & 0 \\ 0.02167 & 0.22931 & -0.00719 & 0 \\ 0 & 0 & 0.00384 & -0.01017 \end{bmatrix}$$

and $\lambda_1 = -0.0029$ (per month) was the crucial eigenvalue that determined the long range half-life of 20 years for all four compartments in the system.

It is not difficult to write a program that forms the cofactor matrix $[C_{ij}]$ of $\lambda_1 \mathbf{I} - \mathbf{A}$. (Many BASICS have a built in determinant function. See page 39.) For $\lambda_1 = -0.0029$, we have

$$\mathbf{C} = 10^{-4} \begin{bmatrix} 3.575 & 1.104 & -0.1031 & 0 \\ 0 & 1.429 & -0.1334 & 0 \\ 0 & 0 & -1.406 & 0 \\ -4.333 & -1.339 & -1.375 & -48.441 \end{bmatrix}$$

and so $\sum_{k=1}^{4} C_{kk} = -44.842 \times 10^{-4}$.

It follows that $\partial\lambda_1/\partial a_{21} = 0.0797$, $\partial\lambda_1/\partial a_{31} = 0.0797$, $\partial\lambda_1/\partial a_{32} = 0.03187$, $\partial\lambda_1/\partial a_{03} = -0.03135$, $\partial\lambda_1/\partial a_{43} = -0.000692$, and $\partial\lambda_1/\partial a_{14} = -1.08024$. If the transfer coefficients can be in error as much as 5%, then the change in the crucial eigenvalue $\Delta\lambda_1$, can be approximated by

$$E = \frac{\partial\lambda_1}{\partial a_{21}} \Delta a_{21} + \frac{\partial\lambda_1}{\partial a_{31}} \Delta a_{31} + \frac{\partial\lambda_1}{\partial a_{32}} \Delta a_{32} + \frac{\partial\lambda_1}{\partial a_{03}} \Delta a_{03} + \frac{\partial\lambda_1}{\partial a_{43}} \Delta a_{43} + \frac{\partial\lambda_1}{\partial a_{14}} \Delta a_{14}$$

and $|E| \leqslant 0.0012$. The crucial eigenvalue can therefore be as large as -0.0041, and the resulting long-range half-life estimate could be as small as 14 years. This computation illustrates the fact that a high degree of accuracy is needed for the transfer coefficients in order to make an accurate half-life prediction.

Sensitivity formulae for difference equations of the form $\mathbf{X}(t+1) = \mathbf{A}\,\mathbf{X}(t) + \mathbf{B}$ are developed in the exercise sections. In particular, eigenvalue sensitivity formulas for Leslie matrices can be derived which are far simpler than the cofactor formulas for $\partial\lambda_1/\partial a_{ij}$ that we have presented. Note, however, that when the entries in the matrix $\mathbf{A}$ are independent, the formula

$$\frac{\partial\lambda_1}{\partial a_{ij}} = -C_{ij} / \sum_{k=1}^{n} C_{kk}$$

can still be applied. This is illustrated in our final example.

Example 12.9 In applying the Leslie matrix model to fish populations, it is often extremely difficult to obtain an accurate estimate of the small survival rate S_0 for the newborns. If λ_1 is the dominant eigenvalue for the Leslie matrix shown below, find $\partial\lambda_1/\partial S_0$, the sensitivity of the dominant eigenvalue to the juvenile survival rate S_0.

$$\mathbf{P} = \begin{bmatrix} 0 & 0 & 500 & 500 \\ 0.005 & 0 & 0 & 0 \\ 0 & 0.4 & 0 & 0 \\ 0 & 0 & 0.6 & 0 \end{bmatrix}$$

Solution 12.9 The Leslie matrix **P** has dominant eigenvalue $\lambda_1 = 1.1502$, and the cofactor matrix of $\lambda_1 \mathbf{I} - \mathbf{P}$ is easily seen to be

$$\mathbf{C} = \begin{bmatrix} 1.5217 & 0.0066 & 0.0023 & 0.0012 \\ 350.04 & 1.5217 & 0.5292 & 0.2760 \\ 1006.54 & 4.3755 & 1.5217 & 0.7938 \\ 661.48 & 2.8755 & 1 & 0.5217 \end{bmatrix}$$

and so $\sum_{k=1}^{4} C_{kk} = 5.0867$. Hence $\partial\lambda_1/\partial S_0 = -C_{21}/5.0867 = -68.815$. All other eigenvalue sensitivities are relatively small. For example, $\partial\lambda_1/\partial S_1 = 0.8602$. If the error in estimating S_0 were as large as 0.002, then the corresponding error for the dominant eigenvalue, $\Delta\lambda_1$, is about 0.138.

EXERCISES

Part A

1 For the continuous infusion model on p. 115, the amount of tracer satisfies the differential equation $\dot{x} = -\lambda x + I$.

(a) Let $\hat{x}$ = limit $x(t)$ and find expressions for the sensitivities $\partial\hat{x}/\partial\lambda$ and $\partial\hat{x}/\partial I$.

(b) If $I = 100$ (mg/minute) and the error in estimating λ can be as great as 5%, estimate the maximum percentage error for $\hat{x}$.

2 For the glucose tolerance test in Example 8.5, $I = 300$ mg/minute, $\hat{c} = c_\infty = 139.2$ mg/dl and $\lambda = 0.05$. If $|\Delta\hat{c}| \leqslant 1$ and $|\Delta\lambda| \leqslant 0.01$, estimate the maximum error in computing the volume of distribution V.

3 In Exercise 11 of Chapter 8, we presented a drug–dose scheme in which single doses of x_0 mg were given every τ minutes and individual doses decayed according to the law $x(t) = x_0 \exp(-\lambda t)$.

(a) If M is the maximum drug level, compute the sensitivity $\partial M/\partial\lambda$.
(b) If $x_0 = 5$ mg, $\tau = 60$ minutes, and the half-title of the drug is estimated to be 90 minutes, estimate ΔM if the error in measuring the half-life is no larger than 10 minutes.

4 For the oral glucose tolerance test in Exercise 3, Chapter 8, compare the two sensitivities $\partial \mathbf{X}/\partial\lambda_1$ and $\partial \mathbf{X}/\partial\lambda_2$.

In the following two exercises, use the 4th order Runge–Kutta method to find $\mathbf{S}_{ij} = \partial \mathbf{X}/\partial a_{ij}$ or $\hat{\mathbf{S}}_{ij} = \partial \hat{\mathbf{X}}/\partial a_{ij}$.

5 In Exercise 5, Chapter 1, we presented a linear compartmental model for nutrient cycling in a simple aquatic food chain.

(a) Compute $\partial \mathbf{X}/\partial a_{ij}$ at time $t = 12$ hours for each of the transfer coefficients. Identify those transfer coefficients with high sensitivities.
(b) Find $\partial \hat{\mathbf{X}}/\partial a_{ij}$ for each of the transfer coefficients. If these coefficients can be in error by as much as 10%, estimate the maximum error for $\hat{\mathbf{X}}$.

6 For the non-homogeneous linear system in Exercise 13, Chapter 7, compute each of the sensitivities $\partial \hat{\mathbf{X}}/\partial a_{12}$, $\partial \hat{\mathbf{X}}/\partial a_{21}$, $\partial \hat{\mathbf{X}}/\partial a_{32}$, and $\partial \hat{\mathbf{X}}/\partial a_{13}$.

7 For the DDT plant spraying model in Exercise 7, Chapter 1, we estimated that 60% of the DDT was intercepted by the plants when they were first sprayed. In general, if α is the percentage intercepted, the initial condition takes the form $\mathbf{X}(0) = x_0[\alpha, 1 - \alpha]$. Find $\partial \mathbf{X}/\partial\alpha$ at times 2, 4, 6, 8, and 10 months for $\alpha = 0.6$ and $x_0 = 1$.

8 For the Aleut ecosystem model, compute and compare $\partial \hat{x}_2/\partial a_{21}$ and $\partial \hat{x}_6/\partial a_{81}$ to assess the long range effects of atmospheric perturbations on the food sources for the Aleuts.

9 If $\mathbf{A}$ is the transfer matrix shown below,

$$\mathbf{A} = \begin{bmatrix} -0.04 & 0.02 & 0.06 \\ 0.04 & -0.03 & 0.04 \\ 0 & 0.01 & -0.10 \end{bmatrix}$$

(a) Find the crucial eigenvalue λ_1.
(b) Compute $\partial\lambda_1/\partial a_{ij}$ for the various transfer coefficients a_{ij}.

10 For the transfer matrix in Exercise 10, Chapter 1, find $\partial\lambda_1/\partial a_{ij}$ where λ_1 is the crucial eigenvalue of $\mathbf{A}$. If the transfer coefficients can be in error by as much as 10%, estimate the maximum error for λ_1.

11 The Leslie matrix for a salmon population is given in Exercise 4, Chapter 4. If λ_1 is the dominant eigenvalue, compute $\partial\lambda_1/\partial S_0$ and $\partial\lambda_1/\partial F_3$.

12 For the bobcat population in Exercise 5, Chapter 5, the average female litter size is $M = 1.4$ while the adult survival rate is 0.67. The juvenile survival rate, however, depends strongly on the prey density and will be left unspecified. If λ_1 is the dominant eigenvalue for the Leslie matrix, compute $\partial\lambda_1/\partial S_0$, $\partial\lambda_1/\partial S_1$, and $\partial\lambda_1/\partial M$ in terms of S_0.

Part B

13 Suppose that $\mathbf{X}(t)$ satisfies the difference equation $\mathbf{X}(t+1) = \mathbf{AX}(t) + \mathbf{B}$ and let $\mathbf{S}_{ij} = \partial\mathbf{X}/\partial a_{ij}$. Assuming that the entries in matrices $\mathbf{A}$ and $\mathbf{B}$ are independent, show that $\mathbf{S}_{ij}$ satisfies the difference equation

$$\mathbf{S}_{ij}(t+1) = \mathbf{AS}_{ij}(t) + \mathbf{Z}_{ij}\,\mathbf{X}(t)$$

where $\mathbf{S}_{ij}(0) = \mathbf{0}$.

14 Let $\mathbf{Y}(t) = [\mathbf{X}(t) \,\vdots\, \mathbf{S}_{ij}(t)]$, where $\mathbf{X}(t)$ and $\mathbf{S}_{ij}(t)$ are as in Exercise 13. Show that

$$\mathbf{Y}(t+1) = \left[\begin{array}{c:c} \mathbf{A} & \mathbf{0} \\ \hdashline \mathbf{Z}_{ij} & \mathbf{A} \end{array}\right] \mathbf{Y}(t)$$

where $\mathbf{Y}(0) = [\mathbf{X}_0 \,\vdots\, \mathbf{0}]$.

15 A Leslie matrix for a swan population is given in Exercise 7, Chapter 5. If $\mathbf{X}(0) = [0, 0, 0, 2]$, use the result of Exercise 14 to compute $\partial\mathbf{X}/\partial F_3$ and $\partial\mathbf{X}/\partial S_3$ at time $t = 10$ years.

FURTHER READING

Peter H. Astor, Bernard C. Patten, Gerald N. Estberg, 'The Sensitivity Substructure of Ecosystems', in *Systems Analysis and Simulation in Ecology*, Volume IV, Academic Press (N.Y.), 1976, pages 390–430.

Answers and Hints to Odd-Numbered Exercises

CHAPTER 1

3 (a) $\begin{bmatrix} -0.2 & 0.05 & 0.25 \\ 0.2 & -0.35 & 0 \\ 0 & 0.3 & -0.25 \end{bmatrix}$ (b) (daily) $\mathbf{X}(1) = [112.5, 182.5, 135]$

(c) (hourly) $\mathbf{X}(1) = [115.86, 192.07, 122.07]$
(d) (continuous transfers) $\mathbf{X}(1) = [115.18, 194.33, 120.48]$

5 (a) $\begin{bmatrix} -0.12 & 0.02 & 0 \\ 0.06 & -0.03 & 0.05 \\ 0.06 & 0.01 & -0.05 \end{bmatrix}$ (b)

t	$x_1(t)$	$x_2(t)$	$x_3(t)$
1	1.857	97.126	1.017
2	3.452	94.490	2.059
3	4.818	92.070	3.111
4	5.987	89.849	4.164
5	6.984	87.809	5.208
6	7.831	85.934	6.235

(c) $\hat{\mathbf{X}} = [10.638, 63.830, 25.532]$

7 (a) $\begin{bmatrix} -0.25 & 0.02 \\ 0.25 & -0.07 \end{bmatrix}$

(b) $x_1(t) = 0.098\, e^{-0.046t} + 0.502\, e^{-0.274t}$
$x_2(t) = 1.00\, e^{-0.046t} - 0.60\, e^{-0.274t}$

9 (a) $x_1(t) = a\, e^{-a_{21}t}$

(b) $x_2(t) = \dfrac{-a\, a_{21}}{a_{32} + a_{52} - a_{21}} \left[e^{-(a_{32}+a_{52})t} - e^{-a_{21}t}\right]$

11

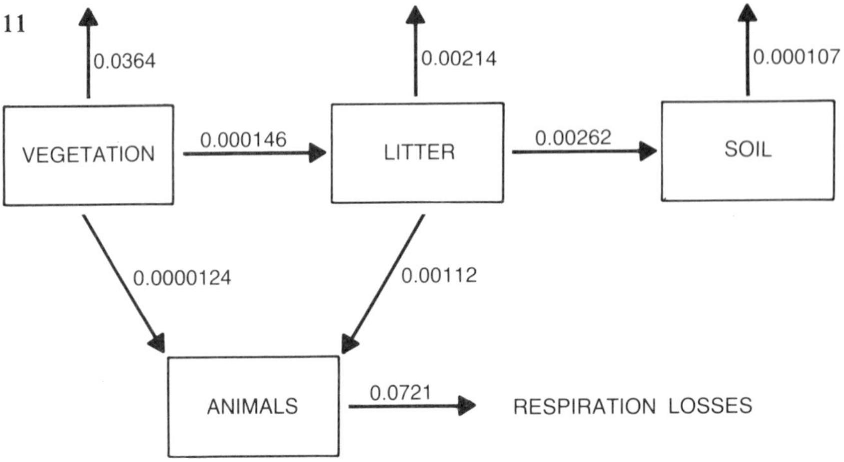

13 (a) reducible (b) irreducible

CHAPTER 2

1 $\mathbf{X}(t) = c_1 \begin{bmatrix} 1 \\ 1 \end{bmatrix} e^{-t} + c_2 \begin{bmatrix} -3 \\ 2 \end{bmatrix} e^{4t}$

3 $\mathbf{X}(t) = -\dfrac{5}{4} \begin{bmatrix} -6 \\ 1 \\ 5 \end{bmatrix} e^{-t} + \dfrac{17}{4} \begin{bmatrix} 2 \\ 1 \\ 1 \end{bmatrix} e^{3t} + 2 \begin{bmatrix} -3 \\ 1 \\ 1 \end{bmatrix} e^{-2t}$

5 $\mathbf{X}(t) = \begin{bmatrix} 5 \\ 0 \\ -5 \end{bmatrix} + \begin{bmatrix} 5 \\ 0 \\ 5 \end{bmatrix} e^{-2t}$ and so $\lim\limits_{t\to\infty} \mathbf{X}(t) = [5, 0, -5]$.

7 $\mathbf{X}(t) = c_1 \begin{bmatrix} 1 \\ 0 \\ -1 \\ 0 \end{bmatrix} + c_2 \begin{bmatrix} 1 \\ -1 \\ 0 \\ 0 \end{bmatrix} + c_3 \begin{bmatrix} -2 \\ 4 \\ 1 \\ 2 \end{bmatrix} 2^t + c_4 \begin{bmatrix} 0 \\ 3 \\ 1 \\ 2 \end{bmatrix} 3^t$

9 $\mathbf{X}(n) = 0$ for $n \geqslant 2$ since $\mathbf{A}^2 = 0$.

11 (a)

$$\mathbf{X}(t) = \begin{bmatrix} 5 \\ 5 \\ 0 \end{bmatrix} + e^{-0.2t} \begin{bmatrix} -5 - t \\ -5 + t \\ 10 \end{bmatrix}$$

(b) $\lim_{t \to \infty} \mathbf{X}(t) = [5, 5, 0]$

(c) $t = 10$

13 *Hint*: Use Theorem 2.5(b) and note that $|e^{\lambda_i t}| = e^{\mathrm{Re}(\lambda_i)t}$. Thus if $\mathrm{Re}(\lambda_i) < 0$, then $\lim_{t \to \infty} |e^{\lambda_i t}| = 0$.

CHAPTER 3

In Exercises 3 to 8, the spectral decomposition $\lambda_1 \mathbf{Z}_1 + \lambda_2 \mathbf{Z}_2 + \ldots + \lambda_n \mathbf{Z}_n$ is given.

3 $$\begin{bmatrix} 2 & 3 \\ 2 & 1 \end{bmatrix} = 4 \begin{bmatrix} 0.6 & 0.6 \\ 0.4 & 0.4 \end{bmatrix} - 1 \begin{bmatrix} 0.4 & -0.6 \\ -0.4 & 0.6 \end{bmatrix}$$

5 $$\begin{bmatrix} 1 & -3 \\ -2 & 2 \end{bmatrix} = 4 \begin{bmatrix} 0.4 & -0.6 \\ -0.4 & 0.6 \end{bmatrix} - 1 \begin{bmatrix} 0.6 & 0.6 \\ 0.4 & 0.4 \end{bmatrix}$$

7 $$\begin{bmatrix} 1 & 2 & 1 \\ 6 & -1 & 0 \\ -1 & -2 & -1 \end{bmatrix} = -4 \begin{bmatrix} 9/28 & -2/7 & -3/28 \\ -9/14 & 4/7 & 3/14 \\ -9/28 & 2/7 & 3/28 \end{bmatrix} + 3 \begin{bmatrix} 16/21 & 4/7 & 4/21 \\ 8/7 & 3/7 & 2/7 \\ -16/21 & -4/7 & -4/21 \end{bmatrix}$$

9 $$\mathbf{X}(t) = \begin{bmatrix} 50 \\ 100 \end{bmatrix} + \begin{bmatrix} 40 \\ -40 \end{bmatrix} e^{-0.6t}$$

11 Eigenvalues are $\lambda = 0$ and $-0.35 \pm 0.1\sqrt{(10)} = -0.35 \pm 0.3162\, i$. The corresponding eigenvectors are $[5, 6, 8]$, $\mathbf{E}$, and $\bar{\mathbf{E}}$, where

$$\mathbf{E} = \begin{bmatrix} -0.98 \ -0.0949i \\ -0.06 \ +0.0949i \\ 0.06 \end{bmatrix}$$

$\hat{\mathbf{X}} = \lim_{t\to\infty} \mathbf{X}(t) = [2.6316, 3.1579, 4.2105]$.

CHAPTER 4

3 (a)

t	Age class 0–1 week	2–3 weeks	4–5 weeks
0	0	100	0
1	2000	0	20
2	0	200	0
3	4000	0	40
4 (8 weeks)	0	400	0

5 $\mathbf{X}_{2n} = (S_0F_1)^{n-1}\,\mathbf{X}_2$ and $\mathbf{X}_{2n+1} = (S_0F_1)^n\,\mathbf{X}_1$

7 $S_0F_1 + F_0 > 1$

9 (a) $\lambda_0 = 1.0028$, $\lambda_1 = -0.7169$, and $\mathbf{S} = [0.7538, 0.1503, 0.0600, 0.0359]$.
(b) $|\lambda_1/\lambda_0| = 0.715$

11 $S_0 > 0.4385$ in order that $\mathbf{P}$ have a dominant eigenvalue > 1.

13 (a) From winter to winter (or summer to summer), the population increases by 212.5%.
(b) The annual spraying program is more effective if done during autumn.

(*Hint*: Compare the dominant eigenvalues of $\mathbf{P}_2\mathbf{P}_1\mathbf{H}$ and $\mathbf{P}_1\mathbf{P}_2\mathbf{H}$ where

$$\mathbf{H} = \begin{bmatrix} 1/2 & 0 & 0 \\ 0 & 2/3 & 0 \\ 0 & 0 & 3/4 \end{bmatrix}$$

15 (a) $\lambda_0 = 0.9989$ and 99.9955% of the population in the stable age distribution is eggs. The vector below gives the proportions in the other age classes:

10^{-7} [212.3, 84.3, 50.6, 40.5, 25.9, 14.8, 8.4, 4.8, 2.7, 1.6, 0.88, 0.50, 0.29, 0.16, 0.09]

(c) The new dominant eigenvalue is $\lambda_0 = 0.9356$. It takes 11 years for the fishable population of 793 000 striped bass to be reduced in half to about 350 000.

17 Assuming that $\mathbf{P}^{-1}$ exists, $\mathbf{X}(0) = (\mathbf{P}^{-1})^n\, \mathbf{X}(n)$.

19 *Hint*: Note that $p_n(\lambda)$ will have a root > 1 provided that $p_n(1) < 0$. Follow the line of attack in the beginning of the proof of Theorem 4.3.

CHAPTER 5

5 (c) $S_0 = 0.236$
(d) $\lambda_0 = 1.974$ or 97.4%/year

7 (a) $x_0(t+1) = (0.96)\,(3.1/2)\, x_3(t+1)$ and $x_3(t+1) = 0.9\, x_2(t) + 0.9\, x_3(t)$.
(b) $\lambda_0 = 1.4327$ and $\mathbf{S} = [0.372, 0.234, 0.127, 0.248]$
(c) The assumption that only one mated pair escaped in 1962 leads to a prediction of 106 swans after 10 years and 150 swans after 11 years. If there were two mated pairs, the population predictions would be doubled. It therefore seems more likely that this population was founded by a single pair.

11 (a) **B** has dominant eigenvalue $\lambda = 1.266629$ and so $h = 0.789497$. The stable size distribution is given by

$$\mathbf{S} = [0.4621, 0.2118, 0.1344, 0.0871, 0.1047]$$

(b)

Time t	Class 0	Class 1	Class 2	Class 3	Class 4	Total harvest
1	1882	980	702	379	342	10 148.07
2	1864	917	661	414	366	10 280.47
3	1890	883	620	421	393	10 368.40
4	1924	874	590	412	417	10 419.19
5	1950	878	572	398	433	10 443.36
6	1965	886	564	385	443	10 450.96
7	1972	894	563	376	447	10 449.95
8	1972	899	565	371	448	10 445.61
9	1970	901	567	369	447	10 440.85
10	1968	902	570	368	446	10 436.91

(c) $\hat{\mathbf{X}}$ = [1963, 900, 571, 370, 444] and the sustainable harvest is approximately 10 425.

CHAPTER 6

3 The dominant eigenvalue of **HP** is 1.13786. Thus the harvesting policy can be implemented and still result in a long-term growth rate of 13.786% per time period.

(b)

Time t	Value of harvest T
1	552
2	306.24
3	912.75
4	515.13
5	1029.69
6	802.99
7	1218.77
8	1134.58
9	1499.46
10	1533.90

5 (a) Since the dominant eigenvalue of **HP** is 0.621618, the population will quickly be driven to extinction.

(b)

Time t	Value of harvest T
1	123.20
2	67.78
3	27.06
4	20.61
5	11.56
6	7.58
7	4.59
8	2.89
9	1.78
10	1.11

Shown in the figure below is the region K defined by the constraints in Problems 7 and 8:

7 The maximum value of the harvest occurs when $\mathbf{X} = [0.625, 0.375]$. We leave the last age class alone and harvest 4/9 of the first age class.

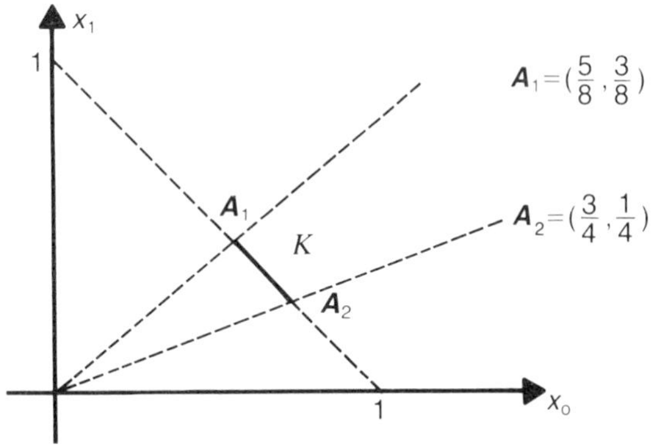

Shown in the figure below is the region K defined by the constraints in Problems 9 and 10:

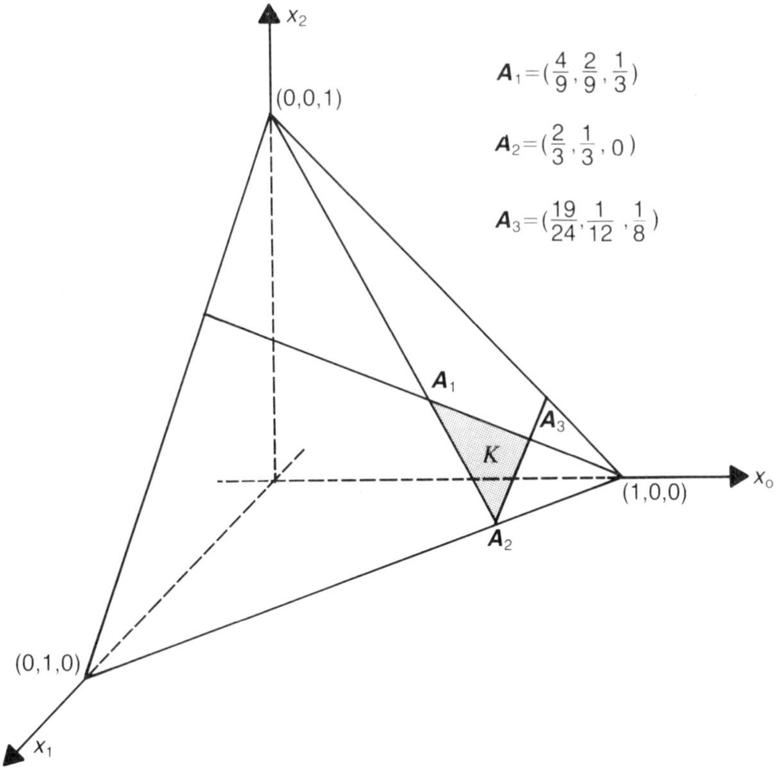

9 The maximum value of the harvest, 0.3125, occurs when $\mathbf{X} = [\frac{19}{24}, \frac{1}{12}, \frac{1}{8}]$. Only the second age class is harvested and 78.95% is taken.

13

Age class	Age distribution	% harvested
0	0.99	52.38%
1	0.00495	0
2	0.00297	0
3	0.00208	0

The maximum value = 2.178 N

15 (a)

Age class	Age distribution	% harvested
0 (fledglings)	0.1	90.855%
1	0.09	0
2	0.081	0
3	0.729	0

(b)

	Age classes			
Time t	0	1	2	3
1	4	25	16	27
2	5	3	23	39
3	8	5	3	55
4	7	7	4	53
5	7	6	6	52
6	7	6	6	52
$\geqslant 7$	7	6	6	52

17 (a)

Size class (girth)	Distribution (before harvest)	% harvested
0 (10–20 in)	0.433	54.09%
1 (21–30 in)	0.150	0
2 (31–40 in)	0.163	0
3 (41–50 in)	0.192	0
4 (51–60 in)	0.062	100%

(b)

Time t	0	1	2	3	4	Harvest
1	2269	903	890	480	93	6 868.04
2	1166	856	922	665	154	7 846.29
3	1107	671	923	802	213	10 173.45
4	1371	553	847	896	257	12 328.57
5	1667	522	752	931	287	13 953.27
6	1904	545	680	919	298	14 733.56
7	2034	593	645	883	294	14 749.12
8	2056	639	643	846	283	14 314.13
9	2008	670	661	820	271	13 759.69
10	1935	681	684	808	262	13 317.18

(c) $\hat{\mathbf{X}} = [1871, 649, 701, 832, 266]$ with sustained harvest of value $\approx 13\,390$.

CHAPTER 7

3 $\mu = 25, K = 50, \omega = \frac{\pi}{12}$, and $t_0 = 8$.

5 $x(t) = c\,e^{-\lambda t} + \frac{K}{\lambda - \alpha} e^{-\alpha t}$, for $\lambda \neq \alpha$

7 $\mathbf{X}_p(t) = \begin{bmatrix} 4 \\ 2.5 \end{bmatrix} e^{-2t}$ *Note*: **A** has spectral decomposition

$$-3\begin{bmatrix} 1.25 & -0.5 \\ 0.625 & -0.25 \end{bmatrix} + 1\begin{bmatrix} -0.25 & 0.5 \\ -0.625 & 1.25 \end{bmatrix}$$

9 $\mathbf{X}_p(t) = \begin{bmatrix} -2 \\ 0 \\ 21 \end{bmatrix}$

11

$$\mathbf{X}_p(t) = \begin{bmatrix} -4 - \frac{1}{47} e^{-3t} \\ 4 + \frac{5}{47} e^{-3t} \\ -12 - 2e^{-t} - \frac{22}{47} e^{-3t} \end{bmatrix}$$

13 $$\mathbf{X}(t) = 3.5714\begin{bmatrix}7\\4\\24\end{bmatrix} + 63.4919\begin{bmatrix}0\\1\\-1\end{bmatrix}e^{-0.35t} - 47.0558\begin{bmatrix}1\\2\\-3\end{bmatrix}e^{-0.25t}$$

$$+\begin{bmatrix}22.0588\\16.3399\\-163.3987\end{bmatrix}e^{-0.08t}$$

Thus $\hat{\mathbf{X}} = \lim_{t\to\infty} \mathbf{X}(t) = [25, 14.2857, 85.7143]$

17 $\mathbf{A}^{-1}$ must exist, that is, 0 cannot be an eigenvalue of $\mathbf{A}$.

19 If $\mathrm{Re}(\lambda_i) < 0$ for each eigenvalue λ_i, then $\hat{\mathbf{X}} = -\mathbf{A}\,\mathbf{Y}_0$, and so $\hat{\mathbf{X}}$ will be independent of $\mathbf{X}(0)$. If 0 is an eigenvalue of $\mathbf{A}$, then the particular solution will be unbounded unless $\mathbf{AX} = \mathbf{Y}_0$ has a non-zero solution.

21 *Hint*: Use Exercise 15, Chapter 3, and Theorem 7.2.

CHAPTER 8

1 Assuming the fasting level is 87.9 mg/100 ml, and ignoring the data point at $t = 180$, you should obtain a value of λ between 0.03 and 0.04. This value is in the normal range.

3 $\dot{x}_1 = -\lambda_1 x_1$ and $\dot{x}_2 = \lambda_1 x_1 - \lambda_2 x_2$ with $x_1(0) = x_0$ and $x_2(0) = 0$.

$$x_2(t) = \frac{x_0\lambda_1}{\lambda_2 - \lambda_1}\,[e^{-\lambda_1 t} - e^{-\lambda_2 t}].$$

5 About 18 years.

7 $c(t) = c_0(1 - e^{-kt})$

9 $x_1(t) = x_0\, e^{-(\lambda_1+\lambda_2)t}$ and $x_2(t) = \dfrac{\lambda_1 x_0}{\lambda_1 + \lambda_2}\,[1 - e^{-(\lambda_1+\lambda_2)t}]$

Only the sum $\lambda_1 + \lambda_2$ can be estimated from blood concentration measurements.

11 Allow at least 1 hour 15 minutes between doses.

13 $F = 3.32$ liters/minute and $V = 0.508$ liters.

15 $\mathbf{X}_p(t) = \left[1, 1 - \dfrac{R}{F_1}, 1 - \dfrac{R}{F_1 + F_2}\right]$ is a particular solution if $I = R$.

CHAPTER 9

1 (a) $x_1(t) = \dfrac{b\,y_0}{m_2 - m_1}\,[e^{-m_1 t} - e^{-m_2 t}]$ and

$$x_2(t) = \frac{y_0}{m_2 - m_1}\,[(a - m_1)\,e^{-m_1 t} + (m_2 - a)\,e^{-m_2 t}]$$

(b) If $\mathbf{X}_1(t)$ is the solution in Theorem 9.2, and $\mathbf{X}_2(t)$ is the solution in part (a) above, then the solution $\mathbf{X}(t)$ with $\mathbf{X}(0) = [x_0, y_0]$ is $\mathbf{X}_1(t) + \mathbf{X}_2(t)$.

3 (a) $x_1(t) = \dfrac{x_0}{a_{12} + a_{21}}\,[a_{12} + a_{21}\,e^{-(a_{12}+a_{21})t}]$

$$x_2(t) = \frac{a_{21}x_0}{a_{12} + a_{21}}\,[1 - e^{-(a_{12}+a_{21})t}]$$

(b) $\hat{\mathbf{X}} = \left[\dfrac{a_{12}x_0}{a_{12} + a_{21}}, \dfrac{a_{21}x_0}{a_{12} + a_{21}}\right]$

(c) $x_1(t)/x_0 = 0.06 + 0.94\,e^{-0.318t}$, where t is in hours.

(d) $a_{21} = 0.301, a_{12} = 0.017$

5 $a_{01} = 0.1359, a_{12} = 0.0706, a_{21} = 0.1435, V_1 = 79.6$ ml, $V_2 = 161.7$ ml

7 When $x_0 = I/m_2$, then $\dot{x}_1 = [(x_0 m_1 - I)(m_1 - a_{12})/(m_2 - m_2)]\,e^{-m_1 t}$ which is either always positive or always negative.

9 (a) $a_{12} = 0.5769, a_{21} = 2.9450, a_{01} = 0.5769$, and $V_1 = 7.563$ liters.

(b) $I = 65.45$ mg/hour $= 1.09$ mg/minute and $x_0 = 358$ mg.

(c) High toxic levels are present for the first 25 minutes. (See page 206 for figure.)

(d) Concentrations are between 15 and 16 μg/ml after about 45 minutes.

11 $x_1(t) = x_0\,e^{-a_{21}t} + \dfrac{R_1}{a_{21}}\,(e^{-a_{21}t} - 1)$ and

$$x_2(t) = \left(x_0 + \frac{R_1}{a_{21}}\right)(1 - e^{-a_{21}t}) - (R_1 + R_2)t$$

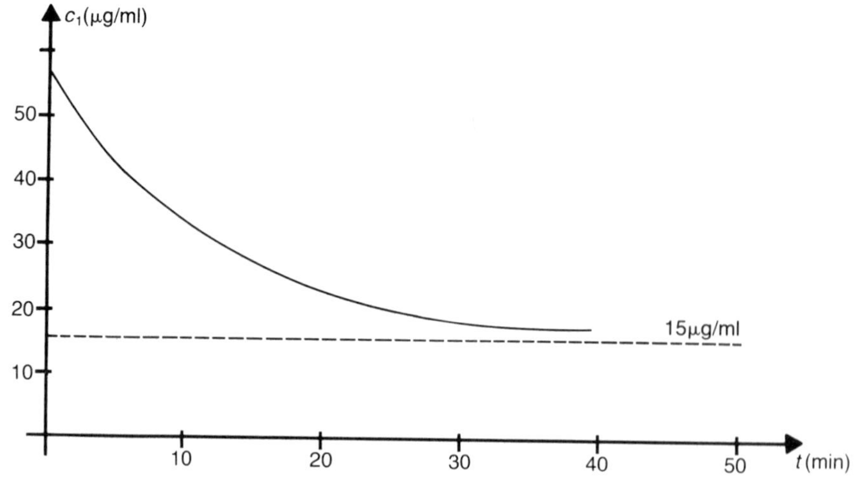

13 Letting $\alpha = a_{03} + a_{13}$ and $D = (a - \alpha)(d - \alpha) - bc$, the particular solution is

$$\mathbf{X}_p(t) = \frac{a_{13}\, x_0}{D} \begin{bmatrix} d - \alpha \\ c \end{bmatrix} e^{-\alpha t}$$

The complementary solution $\mathbf{X}_c(t)$ with $\mathbf{X}_c(0) = -\mathbf{X}_p(0)$ (so that $\mathbf{X}(0) = \mathbf{0}$) is $[x_1(t), x_2(t)]$ where:

$$x_1(t) = \frac{a_{13}\, x_0}{D} \frac{\alpha - d}{m_2 - m_1} [(d - m_1)\, e^{-m_1 t} + (m_2 - d)\, e^{-m_2 t}]$$

$$- \frac{a_{13}\, x_0}{D} \frac{bc}{m_2 - m_1} [e^{-m_1 t} - e^{-m_2 t}] \quad \text{and}$$

$$x_2(t) = \frac{a_{13}\, x_0}{D} \frac{(\alpha - d)c}{m_2 - m_1} [e^{-m_1 t} - e^{-m_2 t}]$$

$$- \frac{a_{13}\, x_0}{D} \frac{c}{m_2 - m_1} [(a - m_1)\, e^{-m_1 t} + (m_2 - a)\, e^{-m_2 t}]$$

(c) Assuming that $c_1(t) = c_1\, e^{-m_1 t} + c_2\, e^{-m_2 t} + c_3\, e^{-\alpha t}$, then $c_3(d - m_1)(\alpha - m_2) = c_1(d - \alpha)(m_2 - m_1)$, from which $d = a_{12}$ can be determined. We can then use the relationships $a = m_1 + m_2 - d$, $bc = ad - m_1 m_2$, $a_{21} = bc/d$, and $a_{01} = a - a_{21}$. Given that $c_1(t) = -12\, e^{-0.15t} + 20.2\, e^{-0.015t} - 8.2\, e^{-0.075t}$, we must identify the $e^{-\alpha t}$ term before we can proceed. Shown in the table below are the three possible cases:

α	a_{12}	a_{21}	a_{01}
0.075	0.0925	0.0482	0.0243
0.15	0.0925	−0.0146	0.0122
0.015	0.0925	0.0109	0.1217

The case $\alpha = 0.15$ must be excluded since $a_{21} < 0$. From the parameters in Example 9.2, the case $\alpha = 0.075$ seems the most likely. Unless a_{13} is known, it is not possible to find a_{03} or V_1.

15 (a) $$x_1(t) = \frac{I}{a_{01}(m_2 - m_1)}[(a_{12} + a_{21} - m_1)\,e^{-m_1 t} + (m_2 - a_{12} - a_{21})\,e^{-m_2 t}]$$

$$x_2(t) = \frac{a_{21} I}{a_{12}a_{01}(m_2 - m_1)}[(a_{12} + a_{21} + a_{01} - m_1)\,e^{-m_1 t} + (m_2 - a_{12} - a_{21} - a_{01})\,e^{-m_2 t}]$$

(b) $a_{01} = 0.0573$, $a_{12} = 0.0598$, $a_{21} = 0.0369$, $V_1 = 7.50$ liters, and $V_2 = 4.63$ liters.

CHAPTER 10

3 Critical points are all of the form $c[9, 5, 3]$. None of these points are locally stable. If $\mathbf{X}(0) = [5, 5, 5]$, then limit $\mathbf{X}(t) = \frac{15}{17}[9, 5, 3]$.

5 The critical points are $[0, 0]$, $[0, 100]$, and $[125, 50]$. Only $[0, 100]$ is locally stable. If $\mathbf{X}(0) = [100, 100]$, then limit $\mathbf{X}(t) = [0, 100]$.

7 The critical points are $[0, 0]$ and $\mathbf{X}_2 = (\ln 375)/3\ [2, 1] \approx [3.9513, 1.9756]$. If $\mathbf{X}(0) = [10, 10]$, then limit $\mathbf{X}(t) = \mathbf{X}_2$. Only $\mathbf{X}_2$ is locally stable.

9 The sole critical point $[645.161, 115.207, 101.382]$ is locally stable, and, if $\mathbf{X}(0) = [0, 0, 0]$, then limit $\mathbf{X}(t) = [645.161, 115.207, 101.382]$.

11 The sole critical point is always locally stable.

13 The period is approximately 45.

15 Using the Runge–Kutta algorithm with $h = 0.1$, 0.01, and 0.001, we obtain (to two correct decimal places) $\mathbf{X}(25) = [0.90, 0.83]$.

CHAPTER 11

1 (a) $0.303\ \text{yr}^{-1}$ (b) respiration loss $= 45\ \text{Kcal/m}^2/\text{yr}$ and $a_{03} = 4.545\ \text{yr}^{-1}$

3 $\lambda_1 = 2.302\ \text{yr}^{-1}$, $\lambda_2 = 1.571\ \text{yr}^{-1}$, and $\lambda_3 = 0.494\ \text{yr}^{-1}$

5 (a)

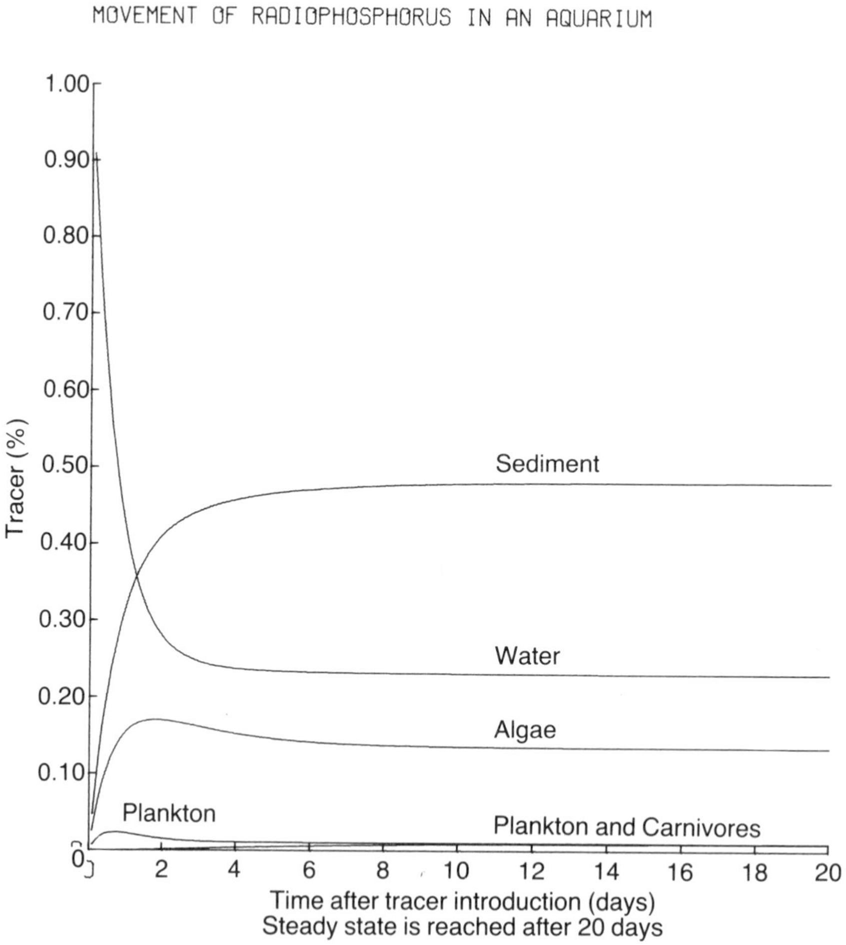

(b) For large t, $\mathbf{X}(t) \approx [13.47, 7.09, 1.09, 1.31, 1.03, 4.81, 48.11, 23.09]$. The largest non-zero eigenvalue is -0.0081.

CHAPTER 12

1 (a) $\partial\hat{x}/\partial\lambda = -I/\lambda^2$ and $\partial\hat{x}/\partial I = 1/\lambda$
(b) 5%

3 (a) $\partial M/\partial\lambda = x_0\tau\, e^{-\lambda\tau}/(1-e^{-\lambda\tau})^2$
(b) $|\Delta M| \leqslant 2190.9\,|\Delta\lambda|$ and $|\Delta\lambda| \leqslant 0.000962$

5 (a)

Transfer coefficient	Phytoplankton	Water	Zooplankton
a_{12}	477.59	−627.48	149.89
a_{21}	−47.74	58.17	−10.43
a_{23}	3.37	54.44	−57.82
a_{32}	−60.25	−658.73	718.99
a_{31}	−51.91	−2.09	54.00

Table entries are state sensitivities at $t = 12$.

(b)

Transfer coefficient	Phytoplankton	Water	Zooplankton
a_{12}	407.24	−746.94	339.70
a_{21}	−67.90	124.94	56.59
a_{23}	54.23	325.94	−380.26
a_{32}	−135.81	−814.85	950.66
a_{31}	−90.54	−11.32	101.86

Table entries are sensitivities at steady state.

7

t	$\partial x_1/\partial\alpha$	$\partial x_2/\partial\alpha$
2	0.5841	−0.5109
4	0.3433	−0.2342
6	0.2037	−0.0796
8	0.1226	0.0048
10	0.0734	0.0491

9 (a) The crucial eigenvalue is $\lambda_1 = -0.07$.
(b) $\partial\lambda_1/\partial a_{21} = -4/3$, $\partial\lambda_1/\partial a_{12} = -1$, $\partial\lambda_1/\partial a_{13} = 0$, $\partial\lambda_1/\partial a_{32} = -1$, and $\partial\lambda_1/\partial a_{23} = 1/3$.

11 $\partial\lambda_1/\partial S_0 = -11.338$ and $\partial\lambda_1/\partial F_3 = -0.0000567$.

Index